T0215822

Perspektiven der Humangeographie

Reihe herausgegeben von
Sybille Bauriedl, Flensburg, Deutschland
Veronika Cummings, Mainz, Deutschland
Martin Doevenspeck, Bayreuth, Deutschland
Florian Dünckmann, Kiel, Deutschland
Johannes Glückler, Heidelberg, Deutschland
Susanne Heeg, Frankfurt, Deutschland
Sebastian Henn, Jena, Deutschland
Judith Miggelbrink, Dresden, Deutschland

In der Schriftenreihe werden Forschungsarbeiten aus allen Schwerpunkten der Humangeographie publiziert. Es werden sowohl qualitativ, wie auch quantitativ ausgerichtete Arbeiten zu aktuellen Fragestellungen des Fachbereichs darin veröffentlicht. Die Reihe ist offen für sehr gute wissenschaftliche Arbeiten, womit sie die Vielfalt und Breite des Forschungsgebietes wiederspiegeln möchte

Weitere Bände in der Reihe http://www.springer.com/series/16066

Tobias Aberle

Entrepreneurship Training in Rural Parts of Bihar/India

Opportunities of Empowering Disadvantaged Youth

 Springer Spektrum

Tobias Aberle
Heidelberg, Germany

Dissertation in Partial Fulfillment of the Requirements for the Academic Degree of Doctor of Philosophy (Dr. phil.) from the University of Education, Heidelberg, submitted by Tobias Aberle, native of Hamburg, Germany in the subject: Geography First reviewer: Prof. Dr. Klaus-Dieter Hupke, second reviewer: Prof. Dr. Alexander Siegmund

Funded in part with a graduate scholarship by the state of Baden-Württemberg (Stipendium im Rahmen der Landesgraduiertenförderung)

ISSN 2524-3381   ISSN 2524-339X   (electronic)
Perspektiven der Humangeographie
ISBN 978-3-658-30007-4   ISBN 978-3-658-30008-1   (eBook)
https://doi.org/10.1007/978-3-658-30008-1

This Springer Spektrum imprint is published by the registered company Springer Fachmedien Wiesbaden GmbH part of Springer Nature.
The registered company address is: Abraham-Lincoln-Str. 46, 65189 Wiesbaden, Germany

Acknowledgements

It is a great privilege for me to be able to devote this dissertation to issues of youth empowerment within the fascinating rural state of Bihar in the North East of India. To have completed it this summer was only possible due to the support and assistance of a range of different people.

I would like to express my special thanks to my supervisor, Prof. Dr. Klaus-Dieter Hupke, for his diligent support and sound advice, and my second supervisor, Prof. Dr. Alexander Siegmund, for his warm and sincere academic support.

Deepest gratitude is owed to Mr. Satyan Mishra, the co-founder and Managing Director of Drishtee, and his team, for their trustful cooperation, and for enabling me to visit the Drishtee training locations in the districts of Madhubani, Bhagalpur and Munger in Bihar, India. I also would like to express my gratitude to Mr. Ashutosh Kumar, for his active support in connecting me with Drishtee.

I would like to express my special thanks to my Indian translators, Mr. Sanjeev Kumar Jha from the Madhubani district, and Mr. Prem Niwas Singh from the Bhagalpur district. I would also like to thank Mrs. Shipra Sapdi for answering any questions related to Hindi-English translations retrospectively, as well as Mr. Jonathan Griffiths for his fast and reliable proofreading.

Finally, this thesis project could not have been finished so smoothly without financial support. I would like to express my thanks to the committee at the Pädagogische Hochschule, Heidelberg for the scholarship awarded to me on the basis of the Landesgraduiertenförderung, as well as the committee at the German Academic Exchange Service for granting me support for my field trip to India.

Contents

Figures and Tables

Abbreviations

ASCI   – Agriculture Skill Council of India
BOP   – Base of the pyramid
BRLP   – Bihar Rural Livelihood Project
CWMG   – Collected works of Mahatma Gandhi
DDUGKY – Deendayal Upadhyaya Gramin Kaushalya Yojana (Skill development program which addresses rural youth from poor backgrounds)
DFID   – Department for International Development (UK)
FAO   – Food and Agriculture Organization (organization of the United Nations)
GEM   – Global Entrepreneurship Monitor
GoB   – Government of Bihar
GoI   – Government of India
GQ   – General research question
ICAR   – Indian Council of Agricultural Research
IFAD   – International Fund for Agricultural Development
ILFS   – Infrastructure Leasing and Financial Services Limited
ILO   – International Labour Organization
KCC   – Kisan Credit Card (a farmers' credit card, which is a product designed by NABARD)
KPMG   – An international auditing company (the letters reflect the names of its founders)
MoF   – Ministry of Finances, Government of India
MoLE   – Ministry of Labour and Employment, Government of India
MoRD   – Ministry of Rural Development, Government of India
MoT   – Ministry of Textiles, Government of India
MSDE   – Ministry of Skill Development and Entrepreneurship, Government of India
MYAS   – Ministry of Youth Affairs and Sports, Government of India
NABARD – National Bank for Agriculture and Rural Development
NGO   – Non-governmental organization
NICRA   – National Innovations on Climate Resilient Agriculture
NPSDE   – National Policy for Skill Development and Entrepreneurship
NRLM   – National Rural Livelihoods Mission
NSDC   – National Skill Development Corporation
NSSO   – National Sample Survey Office (under the Ministry of Statistics and Programme Implementation, Government of India)
NSQF   – National Skill Qualification Framework
OBC   – Other Backward Castes
PMKVY   – Pradhan Mantri Kaushal Vikas Yojana (Skill development initiative of the Prime Minister of India)
RSETI   – Rural Self-Employment Training Institute
RUDSETI – Rural Development and Self-employment Training Institutes

SC	–	Scheduled Castes
SGSY	–	Swarna Javanti Gram Swarozgar Yojna (Self-employment program of the Indian government for the poor)
SHG	–	Self-help group
SQ	–	Specified research question
SSC	–	Sector Skill Councils
ST	–	Scheduled Tribes
VET	–	Vocational Education and Training

German Summary / Deutsche Zusammenfassung

Die berufliche Bildung von Jugendlichen und die Schaffung von Jobmöglichkeiten stellt in Indien, dem Land mit der weltweit größten Jugendbevölkerung, eine der dringlichsten Aufgaben dar. Besonders schwierig gestaltet sich dies jedoch in einem strukturschwachen Staat wie Bihar, in welchem rund 80% der Menschen von der Landwirtschaft abhängen und knapp über 40% – im Vergleich zum indischen Durchschnitt von knapp über 25% – unter der Armutsgrenze leben (GoI 2008; Planning Commission 2009). Infolgedessen hat Bihar nach dem National Sample Survey für 2007–2008 die höchste Abwanderung in andere Teile des Landes (vgl. Bhagat 2016). Dies betrifft insbesondere Jugendliche vom Lande, die für sich keine Zukunft mehr in der Landwirtschaft sehen (vgl. Kumar and Bhagat 2016; Rigg 2006).

Ökonomen verweisen auf die große Bedeutung von Entrepreneurship Development im Zusammenhang mit der Bekämpfung ländlicher Armut (vgl. Ganesh 2010; Hussain et al. 2014; Singer 2006; Yunus et al. 2010). Auch die indische Regierung ergriff Maßnahmen zur Förderung von Selbständigkeit, etwa mit der im Jahr 2011 implementierten National Rural Livelihood Mission, welche nach dem bekannten Vorbild der Grameen Bank in Bangladesh auf der Vergabe von Mikrokrediten an Frauen basiert (vgl. Mehrotra 2016). Weiterhin wurden als Initiative des Ministeriums für ländliche Entwicklung sogenannte Rural Self-Employment Training Institutes in jedem Distrikt Indiens eröffnet, welche jeweils an eine Bank angeschlossen sind und worüber besonders ländliche Jugendliche unter der Armutsgrenze zur Aufnahme einer selbständigen Tätigkeit befähigt werden sollen (RSETI n.a.).

Dies führt zu den zentralen Forschungsfragen, welche Möglichkeiten sich für sozial benachteiligte Jugendliche im ländlichen Bihar über Entrepreneurship Training erschließen, nachhaltige Lebensgrundlagen aufzubauen, welchen strukturellen Limitierungen diese Jugendlichen dabei ausgesetzt sind, und welche Bewältigungsstrategien sie daraufhin entwickeln.

Diesem Thema widmet sich die hier vorliegende Dissertation. Zunächst werden die Feldforschung in Indien sowie die wichtigsten theoretischen und methodischen Grundlagen skizziert. Dann werden die Erkenntnisse der typenbildenden Inhaltsanalyse präsentiert und wichtige Schlussfolgerungen gezogen.

Die Feldforschung in Bihar, Indien

Im Rahmen der Untersuchung wurde im Winter 2015/2016 eine qualitative Befragung unter 25 größtenteils sozial benachteiligten KleinstunternehmerInnen, darunter 14 Jugendlichen im Alter von 18 bis 35 Jahren, in zwei ländlichen Regionen in Bihar durchgeführt. Die Befragten durchliefen ein zwei- bis zwölf-wöchiges Training in den Bereichen Milchwirtschaft, Fischerei, Textilherstellung (Baumwollgarnherstellung, Weben und Nähen) oder Kosmetik (Führung eines Kosmetikladens bzw. eines Schönheitssalons). Der Großteil von ihnen absolvierte das Training bei Drishtee, dem Hauptkooperationspartner für diese Untersuchung, vier Personen bei einem Rural Self-Employment Training Institute und zwei Frauen bei einer kleinen lokalen Nichtregierungsorganisation (NGO), welche nur in Bihar ansässig ist. Drishtee ist ein indisches Sozialunternehmen, welches in den 1990er Jahren gegründet wurde und sich inzwischen zu einer der führenden NGOs im Bereich Social Entrepreneurship entwickelt hat und dessen vielfältige Aktivitäten sich mittlerweile auf 6 000 Dörfer erstrecken (vgl. Drishtee n. a.). Als

Trainingsanbieter ist Drishtee auch Partner der National Skill Development Corporation, welche im Zuge der Realisierung des von der indischen Regierung gesetzten Ziels, bis zum Jahr 2022 eine halbe Milliarde Menschen beruflich zu schulen, ins Leben gerufen wurde (vgl. Planning Commission 2013).

Die jugendlichen EntrepreneurInnen werden in Anlehnung an Giddens' (1984) Strukturationstheorie als handelnde Akteure im Spannungsfeld zwischen Struktur und Handlung betrachtet. Zur Verdeutlichung des strukturellen Rahmens wurde das Sustainable Livelihoods Framework (DFID 1999) zugrunde gelegt, welches sich mit Strukturen, Prozessen und Verwundbarkeitskontexten befasst, unter denen sich Akteure den Zugang zu Lebensgrundlagen wie Human-, Finanz-, Sozial-, Sach- und Naturkapital zu sichern versuchen. Um Jugendliche in ihrem Erwachsenwerden mit ihren persönlichen Motivationen und Wahrnehmungen berücksichtigen zu können, wurde bewusst eine qualitative Forschungsmethode basierend auf einer überschaubaren Anzahl von Leitfadeninterviews gewählt. Dabei erschien ein zweistufiges Auswertungsverfahren als sinnvoll (vgl. Kuckartz 2016). Zunächst wird über eine evaluative Inhaltsanalyse Aufschluss über den individuellen Empowerment-Prozess der untersuchten EntrepreneurInnen gegeben, nämlich bezogen auf die Dimensionen des persönlichen, finanziellen und sozialen Empowerments (vgl. Roy und Saini 2009). Darauf aufbauend wird anhand einer typenbildenden Inhaltsanalyse aufgezeigt, welche unterschiedlichen Strategien zum Aufbau nachhaltiger Lebensgrundlagen Jugendliche in Abhängigkeit von ihrer sozialen Herkunft, von ihrem Bildungsstand und Geschlecht entwickeln.

Die Ergebnisse der typenbildenden Inhaltsanalyse

Aus Gründen der Übersichtlichkeit soll im Rahmen dieser Zusammenfassung nicht auf die erwachsenen, sondern lediglich auf die jugendlichen EntrepreneurInnen eingegangen werden. Im Zuge der Auswertung stellte sich heraus, dass die Jugendlichen sich insbesondere nach zwei Kriterien unterscheiden lassen: Zum einen inwieweit sie sich auf den Ausbau der selbständigen Tätigkeit fokussierten, zum anderen inwieweit sie Eigeninitiative in Bezug auf den Ausbau der selbständigen Tätigkeit entwickelten. Demzufolge lassen sich vier Livelihood-Strategien identifizieren, welche von den drei von Scoones (1998, 4) benannten Livelihood-Strategien, nämlich agrikulturelle Intensivierung/Extensivierung, Livelihood-Diversifizierung und Migration abgeleitet wurden. Auf dieser Grundlage wurden fünf Typen gebildet, welche im Folgenden erläutert werden (vgl. dazu *Figure 26* in der Dissertation).

Die *BusinessoptimiererInnen* (Optimizers) waren stark auf den Ausbau der selbständigen Tätigkeit fokussiert und setzten dabei gleichzeitig auf persönliches Empowerment, indem sie ein unterschiedliches Maß an Eigeninitiative entwickelten, d.h. jegliches proaktives Verhalten jenseits der normalen Routine und bloßen Anwendung von Trainingsinputs, um strukturelle Limitierungen zu überwinden und das Business zu optimieren (vgl. dazu Frese 2009). Dazu gehören z.B. Maßnahmen zur Erlangung eines Kredits, etwa über die freiwillige Vernetzung in einem Farmers Club oder einer Selbsthilfegruppe. Von den 14 jugendlichen Befragten lassen sich drei dieser Gruppe zuordnen. Zwei von ihnen kommen aus armen Verhältnissen. Dies ist sehr bemerkenswert, denn ihnen gelang es, die sie limitierenden strukturellen Rahmenbedingungen stellenweise zu verbessern.

Dem gegenüber stehen in diesem Falle vier *BusinessaufrechterhalterInnen* (Preservers), welche sich ausschließlich reaktiv verhielten, d.h. sie blieben passiv in Bezug auf die sie limitierenden Rahmenbedingungen und hielten ihr Business lediglich am Laufen, anstatt es weiterzuentwickeln. Einer von ihnen stammte aus einer traditionellen Weberfamilie und hatte keinerlei Schulbildung.

Eine weitere Gruppe stellt mit drei EntrepreneurInnen die der *KarriereplanerInnen* (Career Planners) dar, welche eine Bildungsstrategie verfolgten, indem sie nebenbei noch studierten bzw. bereits ein Hochschulstudium abgeschlossen hatten, mit dem Ziel, einmal eine gute Anstellung zu erhalten und die selbständige Tätigkeit dann einzustellen. Manche von ihnen waren auch weiterhin auf den Ausbau der selbständigen Tätigkeit fokussiert. In keinem Fall jedoch stand der gewählte Studiengang in Verbindung mit der selbständigen Tätigkeit. Dieser Trend wird durch die Befragung von 16 Jugendlichen, die sich gerade im Training der gleichen Berufsgruppen befanden, bestätigt, von denen zehn studierten bzw. ihr Studium beendeten. Allerdings wählten nur vier von ihnen das Studium bewusst so, dass sie es auch mit dem Business kombinieren können. Drei dieser vier wählten ein wirtschaftswissenschaftliches Studium.

Zudem lassen sich drei der jungen EntrepreneurInnen als *Diversifizierer* (Diversifiers) einordnen, da sie mehr als einer selbständigen Tätigkeit mit geringfügigem Einkommen gleichzeitig nachgingen, anstatt sich auf eine zu fokussieren und diese zu optimieren. Bei zwei von ihnen stand die weitere Tätigkeit in keinem Zusammenhang mit der selbständigen Tätigkeit, in welcher das Entrepreneurship Training absolviert wurde.

Schließlich lässt sich noch die Gruppe der *Angepassten* (Conformists) identifizieren. Diese Gruppe wurde für die drei EntrepreneurInnen – unter ihnen eine Jugendliche – gebildet, denen es nicht gelang, eine selbständige Tätigkeit aufzunehmen, und die folglich nicht einer der genannten Livelihood-Strategien eindeutig zugeordnet werden konnten. Anstatt proaktiv zu handeln, fügten sie sich den sie einengenden Rahmenbedingungen.

Aus der Betrachtung dieser unterschiedlichen Strategien zum Aufbau nachhaltiger Lebensgrundlagen in Abhängigkeit von sozialer Herkunft, Bildungsstand und Geschlecht können die folgenden Schlussfolgerungen gezogen werden.

Überwindung struktureller Limitierungen

Insgesamt ist festzustellen, dass niemand der Befragten durch die Aufnahme bzw. Weiterentwicklung der selbständigen Tätigkeit einen wesentlichen sozialen Aufstieg erfuhr, etwa durch die Überwindung der Armutsgrenze, und das, obwohl die meisten ihr Business bis zu einem bestimmten Grad weiter ausbauen und somit ihr Einkommen steigern konnten. Ein entscheidender Faktor hierfür ist der Zugang zu einem Kredit. Erstaunlicherweise waren nur vier der 25 Befragten Teil einer Selbsthilfegruppe, obwohl diese Einrichtung in der indischen Gesellschaft ein etabliertes Modell ist, um über einen gemeinsamen Topf an Fördermittel heranzukommen (vgl. Islam und Imam 2011). Einer davon war ein junger Milchbauer, der einen Farmers Club mit 22 Mitgliedern leitete und darüber einen Kredit erhielt. Ansonsten ist es für arme Personen ohne Landbesitz, die ihr in der Regel als Familienunternehmen geführtes Business nur im geringfügigen Umfang betreiben, sehr schwer, über eine Bank einen Kredit zu erhalten.

Mit Blick auf die Bedeutung der Entwicklung von Eigeninitiative für die Überwindung von strukturellen Limitierungen stellt sich die Frage, warum die Mehrheit der Befragten sich diesbezüglich eher passiv-reaktiv verhielt. Zum einen mag das mit der verfügbaren Kapazität sowie mit der persönlichen Motivation zusammenhängen. Die wenigsten fokussierten sich lediglich auf das Business, manche investierten nebenher in Bildung, und einige der Frauen nahmen auch schon mit Mitte zwanzig die Rolle als Mutter ein. Zum anderen vermissten aber viele der EntrepreneurInnen eine aktive Nachbetreuung seitens des Trainingsanbieters. Dadurch fehlte die notwendige Aufklärung bezüglich der Möglichkeiten, an einen Kredit heranzu-kommen, oder der Vorteile, sich sozial zu vernetzen.

Inkohärenz zwischen den persönlichen Zielen der Jugendlichen und der Ausführung der selbständigen Tätigkeit

Bemerkenswert ist der hohe Bildungsgrad einiger der ärmeren jungen EntrepreneurInnen vom Lande, der mitunter bis hin zum Bachelorabschluss reicht. In der Regel streben sie damit prestigeträchtige Jobs an. Ganz oben auf der Beliebtheitsskala stehen dabei Jobs im Staatswesen. Währenddessen haben die erlernten selbständigen Tätigkeiten im Bereich der Landwirtschaft oder des traditionellen Handwerks bei jungen Personen offenbar ein erhebliches Imageproblem. Ein Leiter eines Trainingscenters für Milchviehhaltung merkte an, dass für alle Trainees diese Tätigkeit nur der Plan B sei. Demzufolge führen viele diese Tätigkeiten nur aus, weil sie keine andere Wahl haben. In der Regel folgen sie damit der eigenen Familientradition. So kamen etwa im Bereich der Milchviehhaltung alle Jugendlichen aus Familien mit Milch-kühen.

Dennoch stellt sich die Frage, warum kaum eine oder einer der EntrepreneurInnen in Erwägung zog, das Business mit dem gewählten Studium zu kombinieren, beispielsweise durch die Wahl eines wirtschaftsorientierten Studiengangs, welcher sich direkt auf die selbständige Tätigkeit anwenden ließe. Könnte eine solche Businessprofessionalisierung ein Weg sein, die Tätigkeit auf ein höheres Niveau zu heben und damit eine neue Zukunftsperspektive zu schaffen? Auch wenn das nötige Know-how hierzu vorhanden ist, mögen fehlende finanzielle Mittel dem eine Grenze setzen.

Fazit

Die Befragung unter den EntrepreneurInnen sowie Trainees lässt die Vermutung zu, dass ein Großteil der Jugendlichen aufgrund des hohen Bildungsstands und des Karrierestrebens eigent-lich keine wirkliche Zukunftsperspektive in den hier erfassten Berufsgruppen sieht. Dennoch scheint die Annahme gerechtfertigt, dass nur ein geringer Anteil der Befragten tatsächlich den Einstieg in derartige Karrierejobs schaffen wird. Für die meisten Jugendlichen werden ihre per-sönlichen Aspirationen wohl eine Illusion bleiben. Deshalb bleibt eine Fokussierung auf Entre-preneurship Training in ländlichen Räumen Indiens sehr relevant.

Jedoch erscheint eine Optimierung des bestehenden Trainingsangebots zwingend erfor-derlich. Dieses sollte erstens besser auf die Bedürfnisse junger EntrepreneurInnen insbesondere aus ärmeren Verhältnissen zugeschnitten sein. Angesichts des hohen Bildungsstands vieler die-ser Personen bedeutet dies, qualitativ höherwertige Trainingsangebote von längerer Dauer und im Einklang mit aktuellen technischen Entwicklungen zu schaffen.

Zweitens sollten entsprechende Trainingsangebote auf innovations-basiertes Entrepreneurship ausgerichtet sein (vgl. dazu MSDE 2015, 15). Dabei sollte die Förderung der unternehmerischen Kreativität junger Menschen im Mittelpunkt stehen, was mit einem gezielten Fokus auf die Entwicklung von Eigeninitiative einhergehen muss. Die Untersuchung zeigte jedoch, dass im Zuge der Entwicklung von Eigeninitiative strukturelle Limitierungen nur sehr begrenzt überwunden werden können.

Daher sollte drittens Entrepreneurship Training in einem *Eco-System* eingebunden sein. Dieses besteht nach Prahalad (2014, 13) aus verschiedenen beteiligten Akteuren, wie etwa den Trainingsanbietern, den kreditgebenden Banken, kleinen wie auch großen Unternehmen und nicht zuletzt den staatlichen Institutionen. Auf diesen Akteuren liegt die Hauptverantwortung zur Schaffung besserer Rahmenbedingungen. Auch die indische Regierung verfolgt mit der *National Policy for Skill Development and Entrepreneurship* das Ziel der Schaffung eines „*Eco-Systems* des Empowerments", welches über ein umfangreiches berufliches Trainingsangebot sichergestellt werden soll (MSDE 2015, 11). So ein *Eco-System* ist allerdings nur effektiv, wenn auch junge EntrepreneurInnen aus ärmeren Bevölkerungsschichten im ländlichen Raum darin ernstgenommen werden. Dafür ist es erforderlich, dass seitens aller beteiligten Akteure eine neue Sichtweise auf die entsprechenden EntrepreneurInnen eingenommen wird. Denn „nur wenn wir aufhören, die Armen lediglich als Opfer zu betrachten, und anfangen, sie als robuste und kreative EntrepreneurInnen wahrzunehmen, eröffnet sich eine neue Welt von Möglichkeiten" (Prahalad 2014, 25, eigene Übersetzung).

1. Introduction

1.1 Indian youth and the qualification issue

In India – the country with the world's largest youth population – vocational training and the creation of adequate job opportunities for the young generation constitute one of the country's most urgent tasks. The realization of these tasks becomes extremely staggering in the light of the country's enormous regional disparities. Despite great expertise within the country, a large bulk of the Indian population lives in poor living conditions, being cut off from the economic boom. This is especially true in a state like Bihar, in which 80% of the people depend on agricultural activities and over 40% live below the poverty line (BPL). These statistics stand against the Indian average of just above 25% of the country as a whole who live BPL (GoI 2008; Planning Commission 2009).

Hence, the target group of the present dissertation, the disadvantaged youth sector from rural parts of Bihar, face a threefold set of disadvantages. Firstly, they are geographically disadvantaged, as they grew up in rural parts of Bihar, one of the structurally weakest and poorest states of India. Secondly, they are socioeconomically disadvantaged because of their family status. This often includes educational deprivation. And thirdly, they have a disadvantage in their being *young* entrepreneurs. As members of "youth", these entrepreneurs are still going through a transitional phase into adulthood, and they may consequently still financially depend on their parents and not yet have a strong voice.

However, the high proportion of the child and youth population in India opens a demographic dividend due to a declining dependency ratio[1], which is expected to continue at least until at least the year 2040 (according to United Nations Population Division Projection 2011; see also Mitra and Verick 2013, 10; Mitra and Nagarajan 2005; Navaneetham 2010; Planning Commission 2008/2013). This underlines the urgency for India to empower its young generation through education and training and to generate new job opportunities for them. If this enormous task was neglected, it would "create an army of unemployed youth in the country, leading to a host of socioeconomic problems with wider consequences" (Sankaranarayanan 2011, 56).

In order to face these challenges, the Indian government has defined some ambitious educational goals. The goal of achieving "education for all" has been expanded to the secondary stage. In the area of vocational education, the goal is to qualify half a billion people by the year 2022 (National Policy on Skill Development, MoLE 2009). Moreover, the Indian government has not only formulated ambitious goals but commenced some structural changes as well. In 2009 the "National Skill Development Corporation" was established in a public-private partnership mode to delegate some of the training tasks. In 2014 the "Ministry of Skill Development and Entrepreneurship" was established to bundle together the responsibilities for vocational training and education which used to rest in the hands of 17 separate ministries. How do young people from socially deprived sections of the society benefit from such initiatives, especially those growing up in remote areas, apart from urban agglomerations?

---

[1] The ratio of individuals aged 0-14 and above 65 in comparison to those aged 15-64.

© The Editor(s) (if applicable) and The Author(s), under exclusive
license to Springer Fachmedien Wiesbaden GmbH, part of Springer Nature 2020
T. Aberle, *Entrepreneurship Training in Rural Parts of Bihar/ India*, Perspektiven der Humangeographie, https://doi.org/10.1007/978-3-658-30008-1_1

1.2 The need to focus on a state like Bihar

Taking these regional disparities into account, India can be divided according to economic and demographic aspects into three sets of states (Shanmugam and Moorthi 2010): those with a high population growth rate, but a low average income, those with low population growth and a moderate income, and those with a high income but a moderate population growth rate. The first group is characteristic of the so-called BIMARU-states[2], which refers to the four central Indian states of Bihar, Madhya Pradesh, Rajasthan and Uttar Pradesh. The pointedly chosen similarity of BIMARU with the Hindi word *bimar* (meaning "sick") serves as a reminder of the widespread poverty and weak infrastructure within these states, which host 40% of the Indian population. Particularly backward is the state of Bihar, which is marked by an intergenerational nature of landlessness and poverty (Behera et al. 2013). A major issue in this state, which is dominated by agriculture, is the absence of industry, coupled with a lack of job opportunities in the non-agricultural sector. But the agricultural sector also provides a far from promising future; the agricultural productivity of this state is one of the lowest in India. 93% of the farmers are small and marginal, which is a higher percentage than for all of India (84%). Only 29% of the households own any land and the average landholding size is only one acre (NCEUS 2008). Therefore, in rural parts of Bihar people strive hard to access sources of livelihoods, which is reflected in a high work occupation rate. This often keeps young people from completing their school education or acquiring vocational skills (Mitra and Verick 2013).

As a consequence of these factors, Bihar is the Indian state with the highest rate of people who migrate to other parts of the country (National Sample Survey for 2007–2008, see Bhagat 2016). This concerns especially young people from rural parts of Bihar, who find no future prospects for themselves in the agricultural sector (Kumar and Bhagat 2016; Rigg 2006). Given the very young population in Bihar, which has a median age of 20 years – as opposed to 25 years for all of India (Census of India 2011) – the essential question is how to harness the demographic dividend through qualification initiatives within Bihar, which would open up long-term perspectives for young people to stay in Bihar. This brings the focus to the core subject of this research project, namely entrepreneurship training in rural parts of Bihar. What opportunities can entrepreneurship training provide to empower the disadvantaged youth sector to build up sustainable livelihoods?

1.3 Entrepreneurship training – an opportunity to empower Bihar's disadvantaged youth?

Economists often refer to the great significance of entrepreneurship development in connection with poverty alleviation (Ganesh 2010; Hussain et al. 2014; Singer 2006; Yunus et al. 2010). The Indian government has also taken steps to promote entrepreneurship, for example by launching the National Rural Livelihood Mission in 2011, which is based on the lending of microcredits to women, following the model of the renowned Grameen Bank[3] in Bangladesh

---

[2] BIMARU is an acronym for these four Indian states, and was coined by the Indian demographer Ashish Bose (2005).
[3] The founder of the Grameen Bank, Muhammad Yunus, a Bengalese economist who was awarded the Nobel Prize for Peace in 2006, introduced the concept of micro-financing.

(Mehrotra 2016). In addition, the Indian Government opened under the initiative of the Ministry of Rural Development the so-called Rural Self-employment Training Institutes (RSETI) in every district of the country, with each of them being linked to a bank. The primary goal of this programme is to empower rural youth from below the poverty line to start a self-employed activity (RSETI n.a.). Furthermore, in 2015 the Indian Government formulated the National Policy for Skill Development and Entrepreneurship, which enhances the previous policy for skill development with a specific focus on entrepreneurship. The core objective of the entrepreneurship framework is to "coordinate and strengthen factors essential for growth of entrepreneurship across the country" (MSDE 2015, 13).

Among the most recent initiatives is the Prime Minister's skill certification scheme, called Pradhan Mantri Kaushal Vikas Yoyana (PMKVY). This is the flagship scheme of the Ministry of Skill Development and Entrepreneurship, and is aimed particularly at giving skills to youth (PMKVY 2015). The implementation of this scheme is taking place in partnership with the National Skill Development Corporation. In Bihar a number of private training institutes have become accredited in PMKVY in order to offer short-term entrepreneurship training courses in various agricultural and non-agricultural subjects for young people aged between 18 and 35.

During an extensive field survey between November 2015 and April 2016, various training centres were visited by this researcher in three different districts of Bihar, in order to make contact with self-employed youth who have undergone an entrepreneurship training. In this regard, the Drishtee Foundation, a non-governmental organization with multifaceted services and projects targeted at the sustainable development of rural communities (Drishtee n.a.), was the main cooperation partner. Bihar is one of the key areas of outreach, in which Drishtee launched several centres to provide entrepreneurship training. Guided interviews have been conducted to explore the main research question: how did the entrepreneurship training help disadvantaged youth in rural parts of Bihar, India, to become empowered and thereby to improve their productive livelihoods through engagement in a self-employed activity, and how do these youth cope with structural conditions that enable or constrain them?

This research question runs like a red thread through this whole dissertation, which is constructed in the following way. The theoretical framework, which is introduced in Chapter 2, comprises of mainly three different sources. The first is Anthony Giddens' theory of structuration (1984), which provides a theoretical basis to approaching youth from an action-oriented perspective within the area of tension between agency and structure. Second is the sustainable livelihoods framework, which is useful to understanding the structural conditions within the specific context of rural Bihar, in which self-employed youth gain or increase access to various capital assets. Thirdly, the youth body is considered as productive processors of reality (Hurrelmann 1988/2012; Hurrelmann and Quenzel 2016). In this light, attention is especially drawn to the accomplishment of different developmental tasks as well as the impact of socialization agents, which affect young people's decisions regarding engagement in a self-employed activity and the development of a livelihood strategy.

Chapter 3 will then proceed to break up this general research question into eight specific research questions, which are derived from both the theoretical framework and the collected data. To this end, three supporting hypotheses are formulated, which concern the core

3

assumptions behind the effects of the entrepreneurship training on the agency of young entrepreneurs, and their ability to cope with structural conditions.

Chapter 4 introduces the methodology. It outlines the different steps of the evaluative qualitative text analysis as well as the type-building analysis, and introduces the corresponding category system.

Chapter 5 provides an overview of the scope of the survey and the visited training locations in the two target regions in the state of Bihar, namely the district of Madhubani in the north of Bihar, and the neighbouring districts of Bhagalpur and Munger at the Ganges River.

Chapter 6 presents the findings of the analysis, which reveals stark differences between the entrepreneurs. Some of them developed personal initiative and thus coped with limiting structural conditions very proactively, while others solely behaved in a reactive way. Furthermore, for many of the young entrepreneurs the self-employed activity was only a Plan B, as in the reality they thrived for high-prestige career jobs. Based on these findings, four types of entrepreneurs are created, who represent different livelihood strategies, which the entrepreneurs use to gain access to sustainable livelihoods.

These types lay the basis for the final chapter, Chapter 7, which draws conclusions with respect to the opportunities of empowering disadvantaged youth through entrepreneurship training. In this chapter, recommendations are formulated for the training providers and the societal and political stakeholders involved as to how to improve structural conditions in favour of young entrepreneurs from deprived sections of the society in rural parts of Bihar.

2. Theoretical Framework

"If we stop thinking of the poor as victims or as a burden and start recognising them as
   resilient and creative entrepreneurs and value-conscious consumers,
   a whole new world of opportunity will open up" – C. K. Prahalad

"Mother India has 2.5 billion arms, two billion of those arms are younger than 35.
   It's the youth of India who can help us fulfil our potential" – N. Modi[4]

These quotations outline the scope of the present chapter. Whereas Prahalad (2014, 25), a
renowned American-Indian economist, recognizes the economic potential of the masses of the
poor with this opening statement from his global bestseller, *The Base of the Pyramid*, the main
interest of the present work lies more with the issue of bringing poor people to a place of being
resilient and creative entrepreneurs. India's Prime Minister Narendra Modi pins the hope of the
nation on its youth, speaking not of one billion heads, but two billion arms, in order to
emphasize their practical abilities. This rightly captures the focus of the present chapter, which
deals with the socially disadvantaged youth sector from rural parts of the Indian state of Bihar,
and with the opportunities that exist to empower them through entrepreneurship training.

   Within the present work, the youth sector is not simply approached through economical
lenses as an *asset* in connection with the growth of youth numbers, nor as a *threat* – in case this
demographic dividend would turn into a *demographic disaster* (Mitra and Verick 2013, 1) – nor
even by characterizing them as mere *victims* of their circumstances. Instead, they are, firstly,
approached from an action-oriented perspective (2.1), just as the present chapter is placed
within the context of an action-oriented development geography. Emphasis is placed on the
potential of the individual to change his or her structural conditions. This takes into account
action- and structuration-related theoretical considerations (Giddens 1984). Would these young
entrepreneurs use their capabilities to "make a difference" (Giddens 1984, 14), when confronted
with the harsh conditions of rural Bihar? Will they simply follow their parents' path and do
things as they are used to being done, or will they unleash their entrepreneurial creativity and
come up with new ideas and innovations and transform rural structures (Scoones 2017)?

   Secondly, young entrepreneurs are approached from a pedagogical perspective as
"youth in transition" (2.2). As they transition from school to work, and from childhood into
adulthood, they are considered by paying especial attention to their personal motivations,
aspirations and perspectives within the process of socialization and in connection with different
socialization agents, such as the family and the training provider. In concentrating on these
aspects, the aim is to understand the informal and formal educational and learning processes
related to young people's entrepreneurial activity. Emphasis is especially placed on the
entrepreneurship training as means of empowerment. Does it provide them with a reliable
alternative to build up sustainable livelihoods, when confronted with the temptation to migrate
into other parts of the country and to pursue a career in the government, the business world,

---

[4]   From a speech made by India's Prime Minister Narendra Modi to Indians and people of Indian origin in Sydney,
   Australia, on November 17, 2014 (Hindi-English translation published by NDTV 2014. My own conversion
   of the numbers from crore into the Western Numeral System).

© The Editor(s) (if applicable) and The Author(s), under exclusive
license to Springer Fachmedien Wiesbaden GmbH, part of Springer Nature 2020
T. Aberle, *Entrepreneurship Training in Rural Parts of Bihar/ India*, Perspektiven
der Humangeographie, https://doi.org/10.1007/978-3-658-30008-1_2

academia, and similar industries?

The final section brings all these theoretical considerations together and applies them practically to the specific context of Bihar (2.3). What opportunities does the disadvantaged youth sector have within their specific current line of work (e.g. dairy farming, textiles or beauty and wellness) and their specific region (e.g. the districts of Madhubani, Bhagalpur or Munger) to enhance rural livelihoods through entrepreneurial activities?

2.1 Approaching youth from an action-oriented perspective

The focus of this study is on young entrepreneurs in rural parts of Bihar, and how they manage to cope with enabling or constraining structural conditions. With this general focus in mind, the focal point of research is the area of tension between *agency* and *structure*. This area of tension has been subject to debate within sociology, a debate which has also extended into social geography. There are two opposing positions, which have been described by the social geographer Benno Werlen (1999, 33ff) as *holism* – the belief that, within a society, individuals are determined by their surrounding structure – and *individualism*, the opposite view, that structures can be determined by the actions of individuals within a society. Tröger (2004, 60) points out the relevance of this discussion for the context of development geography and posits the following question: are development problems caused by the affected people themselves, or by the surrounding limiting structures that entangle them? Since the starting point of this research is the youth sector taken as *individuals*, who with their self-employed activity play an *active* role, this analysis will proceed from an action-oriented perspective. Consequently, the core research question is not set down from a holistic point of view – along the lines of, how does the structure determine the actions of young entrepreneurs? – but from an individualistic one, by asking how these young entrepreneurs cope with their surrounding structural conditions. Yet this should not be understood as simply a reduction to adopt one of these extremes. It is rather necessary to connect this work with the theoretical discussion which helps to find a basis to approach the surveyed youth. This will be the main objective in this sub-chapter.

In order to place this work within the field of development geography, 2.1.1 provides an overview of the theoretical discourse within this field from an action-oriented perspective. 2.1.2 then lays down some theoretical foundations by linking this work with Anthony Giddens' theory of structuration, which helps to examine youth within the duality between agency and structure, and how they develop – within this area of tension – personal initiative (2.1.3). Next in 2.1.4, this theoretical framework will be applied practically to the sustainable livelihood approach, which provides an instrument for action-oriented analysis to understand how the disadvantaged youth sector builds up sustainable livelihoods under certain structural conditions. Since the object of research is entrepreneurship training as the pivotal instrument for empowerment, 2.1.5 links the present work with the ongoing debate about the empowerment of the poor which takes place in India. This will build on the previous theoretical considerations concerning an action-oriented approach.

2.1.1 Youth within the context of development geography

The geographical aspect of this research work stems from the theoretical discourse within the field of *development geography*. This chapter begins by placing the present chapter within this sub-discipline. There is no generally accepted definition of the term 'development geography', as it is "diffuse and contradictory" (Müller-Mahn and Verne 2010, 4, my own translation). Therefore, it is necessary to begin by introducing the past and current trends and concepts that have led to an action-oriented development geography. As this cannot be an exhaustive representation, emphasis is placed only on the trends and concepts relevant for the present work. Furthermore, the question will be elaborated as to how youth can be adequately addressed within the field of development geography. This is particularly important as the present chapter looks at youth as entrepreneurs, i.e. as agents for change.

Past trends towards an action-oriented development geography

A paradigmatic shift took place from *geographical research on developing countries*[5] to an action-oriented *geographical development research*, or what is now more commonly termed as *development geography* (Bohle 2011). In order to comprehend this major shift, it is helpful to pin specific focus on the scientific debate as it took place in Germany over the past decades. Already in 1979 Jürgen Blenck postulated that the focal point of geographical research should not be the geographical space in itself – as "there are no problems of the space in itself" (Blenck 1979, 11, my own translation) – but should rather consist of the people, societies and development processes within a certain geographical dimension. Hence, the object of research has shifted from a focus on developing countries towards development itself, which has also implied a shift towards an action-centred orientation. As a result, nation states lost significance, whereas regional and local actors and institutions and their networks gained significance (Bohle 2011). This led to the demand that geographical research should be embedded into the social sciences. New impetus in this direction was given by the geographical working group on development theories, founded by Fred Scholz in 1976 in Göttingen, which had the goal of considering development theories that are based in social sciences within development geography (Scholz 1988).

After this shift the German debate into development geography in the 1970s and 1980s was mainly shaped by two differing theories (Bohle 2011). Firstly, modernization theory argues that endogenous forces within developing countries cause underdevelopment. These forces include an adherence to traditional customs and practices, such as the caste system in India (Behrendt 1969). Secondly, dependency theory argues that exogenous forces perpetuate and obstruct the development process. These forces include the deformation of societal, economical and spatial structures derived from colonial dependencies (Senghaas 1974). Both theories assume that development could be achieved if developing countries follow a "catching-up development path" of industrialization after the model of industrialized countries (Mies 1993, 55; see also Rostow 1960[6]). However, this process turned out to be unsuccessful in the actual practice of

---

[5] The technical term in German is *Entwicklungsländerforschung*, which has more recently changed to *Geographische* Entwicklungsforschung.

[6] A theoretic basis for the belief in a "catching up-development path" stems from the modernization model by the American economist Walt W. Rostow (1960), in which he identified five stages of economic development,

development, and thus led to the realization that these holistic theories have failed, as both internal as well as external factors had to be considered.

Then a paradigm shift took place in the international development debate, which was induced by the realization that the term poverty could not be reduced to its economic dimension, and likewise development could not simply be measured by the expansion of income and wealth. Consequently, with the introduction of the concept of *human development*, which also considers social and political dimensions such as education, health and political participation (UNDP 1990), the people became the focus of development. In light of this shift and in finding new approaches that take account of the complex heterogeneity in developing countries, so-called 'middle range' theories have emerged in the 1990s, which put the focus on local actors within their immediate socio-spatial environment (Bohle 2011). Among middle range theories are vulnerability approaches based on the concept of rural livelihood set out by Chambers (1989), which he designed in order to expand our understanding of poverty and which has been further concretized by Watts and Bohle (1993). These vulnerability approaches have been expanded to include so-called sustainable livelihood approaches by British developmental organizations, such as the Department for International Development (DFID), Care or Oxfam, and have been practically applied within the context of development corporation. Müller-Mahn and Verne (2010) recently criticize middle range theories on the point that they often have a schematic approach, which is inflexible because of the use of predefined categories. This makes it especially difficult to capture broader societal contexts and the role of dynamic change. In addition, these theories are often attuned to the specific demands of development agencies, as it is in the cases of vulnerability and sustainable livelihood approaches.

Since sustainable livelihood approaches have been developed as action-oriented frameworks out of practice, they need to be embedded together with action-theoretical considerations in order to apply them within development geography. Werlen (1997) has linked action-theoretical considerations with Anthony Giddens' theory of structuration within the context of social geography. Not only that, more recent work within development geography, which also has the character of middle range theories (Müller-Mahn 2001; Müller-Mahn and Verne 2010), has combined these action- and structuration-related theoretical considerations with the understanding of Giddens and Werlen, such as Tröger (2004), who analyzed the activities of local actors for food security in Tanzania, or Schlottmann (2007), who carried out a case study on agroforestry in Tanzania. Dörfler et al. (2003) criticize these authors in relation to Giddens' action- and structuration-oriented approaches that they follow a very subject-centered and individualistic methodology. Instead, Dörfler et al. (2003) present an alternative action-oriented approach which links development geography to Bourdieu's (1979) theory of practice. In this criticism, they wish rather to shift the focus onto the relationship between the individual and society and put greater emphasis on the societal level. They claim that this is more appropriate to comprehend the greater social context of such problems.

---

that would lead from a traditional society to a developed one. This work had a great impact on development policies in the 1960s and 1970s.

Embedding the present work within present-day development geography

In order to place the present work within development geography, it is helpful to begin with clarification of the term *development*, which must be understood within the specific context of this research. When it comes to the empowerment of disadvantaged youth in rural parts of Bihar, what kind of understanding of the term *development* will be assumed in order to draw conclusions regarding outcome? In view of the past trends, it is clear that the assumption of a "catching-up-development", which follows the logic of modernization approaches and which implies conventional classifications of the world (e.g. first/third world) with a Euro-centric world view, should be avoided. Within the area of development politics, development is understood as a goal-oriented process aiming at the improvement of conditions (Rauch 2009). Yet this definition is not universally accepted by the research community, as it conveys the impression that development is controllable by development agencies in a manner that would align it with the value-system of the goal makers (Müller-Mahn 2011). Given the fact that this research was facilitated with the help of an entirely Indian social organization, and that it ties in with traditional Indian handicrafts and agricultural activities, it is reasonable to understand development as a goal-oriented process, although this is not enforced from outside but facilitated from within. Therefore, development is oriented at the improvement of living conditions as expressed by the rural entrepreneurs themselves, who voluntarily decided to take up entrepreneurship training courses in order to achieve certain outcomes. Clearly both sides need to be considered: that is, the role of the local individual actors as well as the structural conditions in place, such as the societal and political system with its norms and regulations.

An appropriate understanding of development must also take into account the process of globalization and changing conceptions of space, which have led to a dissociation of conventional classifications of the world within development geography, such as first/third worlds, and countries of the north/south, on the grounds that they do not match the empirical reality (Escobar 1995; Müller-Mahn and Verne 2010; Simon 2003). Instead, Scholz (2002/ 2004) understands development as a fragmented process which creates the parallel existence of rich and poor in confined spaces. He identifies the masses of the poor as an "*ocean of poverty*" (Scholz 2002, 8; my own translation), who make the "*population redundant*"[7] (Scholz 2000, 4) by their exclusion through three means, namely their redundancy as manpower, their insignificance as consumers, and the insignificance of their goods. This perception has been questioned by recent studies such as Kulke and Staffeld (2009), who investigated informal production systems in Dhaka, Bangladesh and who conclude that the poor are without question marginalized, but not excluded from society, as they form the basis of an efficient production system. Moreover, Bronger and Trettin (2008) emphasize the social cohesion of people residing in marginal settlements in Mumbai, India. In connection with the present work, the masses of the rural poor in India, such as the small-scale farmers in marginal areas who make up 86% of all farmers and are considered the "backbone" of the Indian agriculture (Sengupta 2007, 134), also cannot be considered as excluded. Development is therefore understood as a process of

---

[7]   Scholz took the phrasing "population redundant" from the British economist David Ricardo (1772-1823), who dealt with the social effects of the substitution of machinery for human labour, and presumed that on the one hand, it may increase the net revenue of the country, but on the other hand it "may at the same time render the population redundant, and deteriorate the condition of the labourer" (Ricardo 1817, in Kurz 1994, 92-93).

empowering the poor, which results in economic and social inclusion through the active participation of the poor in their development process.

With this understanding of development in mind, the present work can successfully be embedded within present-day development geography, which geographers have described as action-oriented, multi-dimensional and transdisciplinary (Bohle 2011; Müller-Mahn 2011; Rauch 2009). The action-oriented aspect is provided by the basic research question, which puts emphasis on the agency of young entrepreneurs, that is to say the way they cope with structural limitations. In this context action-theoretical reflections are drawn from Giddens' theory of structuration. Even though it may be reasonable to draw on Bourdieu in order to consider the larger societal dimension, this would expand the scope of this research, as further societal or political agents would have to be included in the investigation. However, the methodological problem remains of narrowing down the problem to the specific context on the one hand, and of broadening the perspective on the problem as much as possible to capture the social reality on the other hand. This problem will be resolved through a multi-dimensional approach, which includes a multiple group of actors in the survey, the entrepreneurs, as representatives of the training providers and the village community. The criticism of following too great a subject-oriented or individualistic methodology is counteracted by the transdisciplinary design of our conceptual framework, which includes connections to the educational sciences and even draws on sociological and psychological insights in considering the socialization of youth.

Targeting youth within development geography as agents of change

Young people, or the youth sector collectively, have rarely been specifically targeted within development geography (Mayer 2002b). However, they have become a central policy concern in India and beyond, especially with regard to the skilling mission and issues of unemployment (Planning Commission 2013). In view of accelerating technological development, new educational opportunities and the process of globalization, it seems important to understand how youth nowadays gain access to livelihoods and the role they play in the process of social development.

In his work on youth conflicts and development planning in Sri Lanka, Mayer (2002a; 2002b) addresses the youth sector within the context of development geography and considers young people within their historic and cultural contexts as a relative concept, whilst also taking into account their regional-specific social processes. Instead of defining youth by age or stage of development, he views them within an active process of dealing with their environment. This perspective will also be assumed for the present work, as it underlines the active role which young people play in dealing with structural conditions, and as it recognizes them as agents instead of passive victims (Bell and Payne 2009).

As young people are in a flexible stage of life, in which networking and mobility are becoming increasingly important, newer approaches within development geography, the so-called alternative development theories, also seem appropriate to addressing youth. These approaches consider space not as a clearly defined unit, but as something relational, defined by networks of flows and relationships (Müller-Mahn and Verne 2010). It would lead us off topic to further deepen the key concepts here, such as Castell's liquefaction of space or the space of flows (Castells 2010). But the implication for the present work is that we should not consider the surveyed young entrepreneurs as a static group, but rather see them, as it were, in a flow, a dynamic and changeable stage of life. This has direct implications for their livelihood strategies, as the decision to remain in traditional rural industries or agricultural activities stands in competition with increasing educational opportunities which entice them to thrive for migration and career opportunities outside of their familiar settings.

2.1.2 Anthony Giddens' theory of structuration: seeing youth as agents

Anthony Giddens' (1984) theory of structuration deals with the interaction of agency and structure, and lays an important conceptual foundation for the present work. In the following section, its relevance within social sciences and ways that it can be applied as an analytical framework to approach youth from an action-oriented perspective will be elaborated. To this end, its key elements, namely agency, structure and the duality of structure will be introduced and specifically applied to the activities of youth as rural entrepreneurs.

The relevance of the theory of structuration lies in exemplifying how society is constituted through the recursive actions of individuals. Giddens deals critically with the works of Marx, Durkheim and Weber through the lenses of sociological action-theory and further develops their practical-sociological ideas to construct a far-reaching and interdisciplinary meta-theory within the social sciences (Werlen 2012). His criticisms regarding these classical sociologists relate to the fact that, in his view, these earlier sociologists have not been able to accurately encompass the connection between individual action and the social structure of society. Giddens emphasizes the subjectivity of the individual actor, although he attempts to mediate between the extreme positions of subjectivism and objectivism. The key elements of the theory of structuration are the following.

Agency

Agency places the agent as subject in focus, as one who is not simply determined by structural conditions, but who has the ability to carry out self-determined action. But at the same time, "agency refers not to the intentions people have in doing things but to their capability of doing those things in the first place" (Giddens 1984, 9). In this sense agency involves power and thus can be considered in connection with empowerment.[8] In this way, Giddens describes agents as *knowledgeable,* as they have mutual knowledge that is shared with other actors. This enables actors to interact with each other as competent agents (Giddens 1984, 281). Hence, knowledge has a key role in the way actions are produced by the actors. This can be described through three stages of consciousness (Giddens 1984, 7):

- *Discursive consciousness,* which contains reflexive knowledge. At this level, action is controllable and can be explained. This is where the reflexive monitoring of action takes place. But the vast bulk of mutual knowledge is not accessible at this level of consciousness.
- *Practical consciousness,* which contains the greatest bulk of knowledge. This is practical in character and "inherent in the capability to 'go on' with the routines of social

---

[8] See under 2.1.5.

life" (Giddens 1984, 4). But it cannot be explained if it is asked about. This is the source from where rationalization for action stems.

- *The unconscious*, which is formed through mechanisms of suppression, which means the actor is either unaware of this area of influence, or her awareness of it is distorted. This is the source from where motivation for action stems.

Whereas the line between the unconscious and practical consciousness is fixed, the line between practical and discursive consciousness fluctuates, as actors are potentially able to reason their actions through their knowledgeability. Yet the line can also be moved towards discursive consciousness, for instance through learning (Tröger 2004). This discovery is key with regard to the surveyed entrepreneurs, for whom the following can be concluded so far: the entrepreneurs can be considered as competent actors. In this way, the type of action that matters for the survey is any action they take in connection to their business activity. Now, all of them have undergone an entrepreneurship training course, which leads to the assumption that their level of competence, or the understanding about their profession, has also increased. This means in effect that, through learning which took place during the entrepreneurship training, the line between practical and discursive consciousness has shifted towards discursive consciousness, depending on how much they have learnt.[9]

Structure
The structure represents social conditions which are made up of rules and resources. It is helpful to differentiate between structure and social systems. Structure only exists in an abstract fashion, being "out of space and time" and being "marked by an 'absence of the subject'" (Giddens 1984, 25). Social systems on the other hand, "comprise the situated activities of human agents, reproduced across time and space" (Giddens 1984, 25). Structure is recursively implicated into social systems. On this view, a self-employed activity would be a social system, which is a continuous activity reproduced within the material dimensions of space and time. The structure additionally encompasses rules and resources. Rules are a particularly essential part of practical consciousness and are reflected in the knowledgeability of actors. They can be either signifying, which means that they give meaning to the actions, or normative in nature, which means that they have a sanctioning function regarding the modes of social conduct. Hence, they are important for the legitimation of action. The resources as a second element of structure do not mean material things; instead they are the "structured properties of social systems, drawn upon and reproduced by knowledgeable agents in the course of interaction" (Giddens 1984, 15). Resources function as media through which power is exercised, and thus refer to a sort of transformative capability, which means "the capability to 'make a difference', that is, to exercise some sort of power" (Giddens 1984, 14). In this, actors can draw upon authoritative resources, which refers largely to exerting power through knowledgeability, but which can also mean to

---

[9] It is worth noting that any information provided by the entrepreneurs during the interview is discursive knowledge, which means only the portion of their knowledgeability, which they are able (and willing) to reflect on and about. Giddens points to the problem of the *"double hermeneutic"* (Giddens 1990, 15) that occurs with regard to the understanding of the provided information, as both the surveyed person and the researcher who conducts the interview have their own theory.

generate the control one has over persons or actors, for instance in the form of a labour contract. Not only that, actors can draw upon allocative resources, which means in the case of business entrepreneurs to exercise power to gain access to sustainable livelihoods, such as natural resources, production devices and products.

Duality of structure

The absolutely core assumption of the theory of structuration is that agency and structure cannot be analysed separately, as one does not exist without the other. Every action is embedded within a structure and without structure there is no action (Giddens 1984). They do not coexist in a dualistic relationship, but are two different sides of one thing: social practice. In order to describe the interplay between agency and structure, Giddens uses the analogy of language, and differentiates between language itself and speaking as the use of language. A language is made up of rules and regulations, and exists detached from time and space. But every time language is spoken, it is reproduced through space and time. It is the same with structure, which exists detached from time and space, but becomes manifest through actions. The dualistic nature of structure is that structure determines social actions, but at the same time actions determine structure, or, in other words, "structure is both medium and outcome of the reproduction of practices" (Giddens 1979, 5). As much as a language determines the way that it is used by people, it is also the product of the way it is used, which means that the use of it changes the language, e.g. the vocabulary, over the course of time. Now Giddens emphasizes that "routinized practices are the prime expression of the duality of structure in respect of the continuity of social life" (Giddens 1984, 282). This is consequently also the case for self-employed activities.

At this point, another important aspect must be noted. Giddens considers human action as a "continuous flow of conduct", or a sequence of actions (Giddens 1984, 3). This means that the flow can be determined by actions, or in other words the capability to act differently. This brings a dynamic element into the analysis which has important implications for the activities of self-employed actors. It is reasonable to assume that such activities play an active part in determining which direction this continuous flow of conduct goes, e.g. by taking various management decisions which are within their scope of action (for instance: Do I expand my business and apply for a credit? Do I purchase more cows? Do I follow a strategy of agricultural diversification with different cultivations?). In any case, it is important to notice that actions have "unintended consequences", which in turn are the "unacknowledged conditions of further acts" (Giddens 1984, 8). The aspect of seeing human action as flow is especially relevant when considering the youth sector, who are in a very dynamic stage of life, as they have to resolve important decisions regarding their education and professional career or marriage. In the case of young entrepreneurs, the more fundamental decision which needs to be resolved is whether they remain in the field of their business in which they took the entrepreneurship training, or whether they escape from it and strive for a professional career.

But an important question arises regarding the choices that young entrepreneurs from disadvantaged social backgrounds have. To what extent are they able to act freely? How restrictive are their structural conditions? To what degree is the flow of conduct determined by their actions or by the structural conditions? It is certainly impossible to resolve this question to the last detail. But key to this analysis is the assumption that the disadvantaged youth sector could become empowered through entrepreneurship training and the assistance of the training provider to change their structural conditions. Consequently, the degree to which the entrepreneurs make use of, or cope with, their structural conditions depends on their level of empowerment. The word *empowerment* includes within it the word *power*. Robson et al. (2007) refer to the mutual contingency of power and agency, which means that there is no agency without power, and there is always an agent needed in order to exert power. According to the Merriam-Webster Dictionary, these two concepts are by definition interrelated, insofar as the agent is defined as "one that acts or exerts power" (Merriam-Webster 2017). Hence, empowerment means to enable the actor to exert power through the means of business activities, by drawing upon structural elements in the form of rules and resources.

Klocker (2007) advanced Giddens' theory of structuration by drawing on Foucault's analysis into the connectivity between power relations and subjectivity, differentiating between *thick* and *thin* agency. She describes thin agency as "decisions and everyday actions that are carried out within highly restrictive contexts, characterized by few viable alternatives", and thick agency as "having the latitude to act within a broad range of options" (Klocker 2007, 85).[10] This distinction is pertinent with regard to very diverse entrepreneurs, who have intentionally been selected from different social and educational backgrounds. There are some entrepreneurs, the earlier-mentioned three types of disadvantaged youth from rural parts of Bihar (see 1.1), for whom a highly restrictive context will be assumed, and therefore thy will be said to assume a thin agency. On the other hand, there are other entrepreneurs from wealthier backgrounds, for whom a thicker agency will be assumed. According to Klocker, the "thickeners" or "thinners" of the agency of individuals can be "structures, contexts, and relationships, (…) by constraining or expanding their range of viable choices" (Klocker 2007, 85).

The following figure illustrates the empowerment-process of entrepreneurs based on Giddens' theory of structuration. Through the learning process, which takes place during the entrepreneurship training, the degree of discursive consciousness of the competent actor increases. As a concomitant effect of the learning, the amount of reflexive knowledge, which allows for the reflexive monitoring of action, increases as well. Now, action is always embedded within structure, which is both the medium and outcome of social practice. It will be assumed that it is possible to apply reflexive knowledge, which has increased through the entrepreneurship training, to the self-employed activity in such a way that the agency would *thicken* over time, and thus the capability to cope with structural limitations would increase as well.[11]

---

[10] Klocker uses *thin* and *thick* agency in the context of young female child labourers in Tanzania, and highlights the fact that there is still an ability to act, even in the face of high pressures (Klocker 2007). Even though the situation of the surveyed entrepreneurs in Bihar is different, this distinction seems applicable with respect to the significant social disparities.

[11] In order to understand young people's agency, a "multiplicity of factors" (Klocker 2007, 85) would have to be considered (which will be done in 2.1.3, 2.1.4 and 2.1.5 below). Hence, this assumption cannot automatically be expected to be true, and therefore it is only formulated as an ideal-typical assumption, highlighting a *possibility* based on the theoretical back-up. As a consequence of this, the hypothesis will be further developed under 3.2 in consideration with the development of personal initiative (2.1.3) and livelihood strategies (2.1.4.2).

Figure 1: The process of empowerment in view of Giddens' theory of structuration

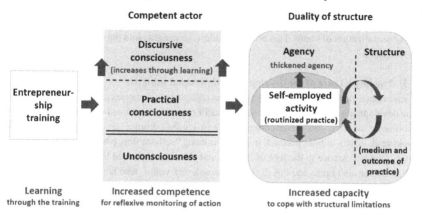

Source: Author's own image, based on Giddens (1984) and expanded by the references to the learning effect (Tröger 2004) and the thickened agency (Klocker 2007).

Criticisms

Dörfler et al. (2003) criticize action-theoretical approaches which underlie the assumption that development problems could be understood and its dependency on local actors and how they deal with various structural constraints. In response, they look to detach themselves from this perceived "methodological individualism" and draw on Bourdieu's theory of practice, in order to build the bridge between objective societal structures and social practices (Dörfler et al. 2003, 15, my own translation, see also Deffner and Haferburg 2014). This concern must be taken seriously. However, in response, it must be noted that this is not a criticism of Giddens' theory of structuration as such, but a criticism regarding the way it is applied in various action-oriented approaches within the context of development geography. Therefore, this needs to be considered in its application to the sustainable livelihoods framework, which will be done in the following way. The actions of youth as individual subjects cannot be simply taken as free and rational, as livelihood approaches may imply, but need to be considered within the duality of structure. The challenge is thus to find ways to represent the following dynamic aspects within a statically designed framework simultaneously:

- The continuous flow of social practices, which is reflected in the constant reproduction of practices through the interplay of agency and structure; and
- The learning process that was initiated by the entrepreneurship training, which must be seen in considering the different levels of consciousness, and the line between the practical and discursive consciousness, which shifts through learning.

This multi-perspective analytic approach, with entrepreneurs viewed as the subject of research and the training providers/entrepreneurship training as the object of research, prevents one from becoming too fixed on either one of the extremes, subjectivism or objectivism. During the field

15

survey, first-hand information was gathered from both sides. Furthermore, an interdisciplinary approach is chosen, which – in addition to action-oriented considerations from the perspective of development geography – takes into account sociological considerations on youth transitions and the socialization of youth. This provides additional insights into the structural set-up of the entrepreneurs.

2.1.3 The importance of the development of personal initiative

An important aspect to shed light on the question regarding free choices and thin or thick agency is the development of personal initiative. Michael Frese, who examined the personal initiative-performance of rural and urban entrepreneurs in Uganda from a psychological perspective (Rooks et al. 2016), defines personal initiative as being "self-starting, [showing] long-term pro-activity, and persistence in the face of barriers and obstacles that need to be overcome" (Frese 2009, 444, see also Frese and Fay 2000). In considering youth, who are in a transitional stage and have to make decisions within the area of tension between agency and structure, it seems appropriate to consider these psychological insights of Frese, who highlights the significance of action within the context of personal initiative by proposing that "there is only one pathway to success and that is through actions" (Frese 2000, 162). In light of the present chapter, the question is whether the entrepreneurs develop personal initiative regarding their constraining or enabling structural conditions, which would be reflected in any kind of proactive behaviour to run the business better, to engage socially, to avail support, and the like.

The contrary of these proactive behaviour patterns would be to act in a very reactive way. Frese considers personal initiative in opposition to reactive strategies, and explains that "the reactive strategy responses to environmental demands, is not proactive, and does not imply any self-starting" (Frese 2000, 164-165). Through his study of micro-entrepreneurs in Africa, he concluded that both personal initiative and reactive strategies are highly related to success. Thus, it can be assumed that, for the surveyed entrepreneurs, the consideration of both reactive and proactive behaviour patterns is crucial for gaining a better understanding of how they cope with limiting structural conditions.[12] At this point, an important remark of Giddens regarding passivity must be taken into account, who states that "to 'have no choice' does not mean that action has been replaced by reaction (in the way in which a person blinks when a rapid movement is made near the eyes)" (Giddens 1984, 15). This means that reactive behaviour still implies active action in the sense of having taken a conscious (or unconscious) decision to not do anything about certain structural conditions, such as to become an active member of a self-help group in order to avail financial support for the business.

2.1.4 Youth and sustainable rural livelihoods

In the following section, the previous structural-theoretical considerations will be applied practically within the context of rural activities, with the aim of achieving safety and a sustainable livelihood. Sustainable livelihood approaches have been developed as practical tools within the area of development politics (Carney et al. 1999, Knutsson 2006). First, it will be comprehended how such frameworks have evolved, and then a specific livelihood

---

[12] This assumption will be revisited under 3.2 and specified as a hypothesis for the analysis.

framework will be introduced in detail, which provides a basis for the present analysis. In this respect, youth will stand in focus of the considerations regarding the access to sustainable livelihoods and the development of a livelihood strategy.

2.1.4.1 The development of sustainable livelihood concepts

The practicability of sustainable livelihood concepts, which are a further development of the vulnerability approach (Krüger 2003), lies in the fact that they focus on particularly vulnerable individuals at the household level in rural living environments, and they systematically outline their access to sustainable livelihoods within an area of tension between agency and structure. The initial impulse for this angle was given by the concept of sustainable livelihood introduced by Robert Chambers and Gordon Conway (1991), who combine the ideas of capability, equity and sustainability in the following way:

- *Capability* is derived from the capability approach of Amartya Sen (1987). Sen uses the term with a wide purview to encompass any capability of an individual or household to do something or be something. This not only means to be able to achieve something, but also to have the freedom to choose among various opportunities, e.g. to choose a certain style of life. A *livelihood capability* includes "being able to cope with stress and shocks, and being able to find and make use of livelihood opportunities" (Chambers and Conway 1991, 4).
- *Equity* refers to the enhancement of living conditions of the most deprived groups of people, which can be achieved by an equal distribution of assets, capabilities and opportunities.
- *Sustainability* within the context of livelihoods is used "in a more focussed manner to mean the ability to maintain and improve livelihoods while maintaining or enhancing the local and global assets and capabilities on which livelihoods depend" (Chambers and Conway 1991, 5).

Various development organizations in the United Kingdom, such as Care, Oxfam, the UK Department for International Development (DFID) and even the United Nations Development Program (UNDP), further developed this concept to apply to policy-oriented sustainable livelihoods frameworks (Knutsson 2006). For the present study, the sustainable livelihoods framework of the DFID seems very suitable as an analytical framework, in order to exemplify the structural conditions under which rural youth in Bihar gain access to sustainable livelihoods and what type of livelihood strategies they develop.

2.1.4.2 The sustainable livelihoods framework of the Department for International Development (UK)

The sustainable livelihoods framework of the DFID (1999, see Appendix X) was adjusted from a framework for analysis developed by Ian Scoones (1998). At first, a definition of sustainable livelihoods will be presented, and then it will be applied to the sustainable livelihoods framework.

Sustainable livelihoods

The following DFID-definition of sustainable livelihoods was adopted, except for some minor changes from a definition proposed by Scoones, which he modified from the original definition of Chambers and Cornway (1991): "A livelihood comprises the capabilities, assets (including both material and social resources), and activities required for a means of living. A livelihood is sustainable when it can cope with and recover from stresses and shocks and maintain or enhance its capabilities and assets both now and in the future, while not undermining the natural resource base" (DFID-definition in Krantz 2001, 3).

How does this definition appear in the context of self-employed youth in rural parts of Bihar? To begin with, a closer look is taken at three elements of which a livelihood comprises: capabilities, activities and assets. As already mentioned, capabilities include the ability of being able to find and make use of livelihood opportunities. This is obviously something that first has to be achieved, and therefore involves a dimension of empowerment, especially in considering youth from poor rural backgrounds. What the surveyed entrepreneurs have in common – most of whom are at an age in which they are officially considered as youth – is that they used the entrepreneurship training as a tool of empowerment in order to be able to find and make use of livelihood opportunities. The second element are activities, more specifically the self-employed activities in the different areas in which they have been trained. The third element, assets, is reflected in the sustainable livelihoods framework as consisting of (the following list after Scoones 1998, Krüger 2003 is non-exhaustive):

- Human capital: knowledge and skills (therefore education and training are important for their acquisition), health;
- Financial capital: cash, savings, credits, which may have been granted from banks;
- Natural capital: land, soil, water, fishpond, biodiversity;
- Physical capital: basic infrastructure, livestock (e.g. cows), production equipment, technology; and
- Social capital: social networks (e.g. membership in a self-help group), social claims, and social relations.

The second part of this definition relates to the aspect of sustainability: a livelihood must be able to cope with shocks and stresses, which are unplanned events that suddenly may occur, such as natural disasters, like a flood or drought. Even though these are important aspects, they have no relevance for the research question, and therefore will not be considered in further detail. Although the surveyed entrepreneurs were exposed to many structural constraints, the time and places have been chosen in such a manner that it was very likely to meet them not in any moment of shock or stress, but in a relatively stable condition, in which they were able to follow their daily routine. The next aspect of a sustainable livelihood is of greater relevance, namely the ability to maintain or enhance one's capabilities and assets both now and in the future. This involves a dynamic element. It will be assumed that through the entrepreneurship training, the capabilities have increased, and if these acquired skills and knowledge are successfully applied to the self-employed activity, assets will also become enhanced.

18

Application to the sustainable livelihoods framework of the DFID

Now, the sustainable livelihoods framework considers access to these livelihood assets under certain structural conditions: on the one hand the context of vulnerability, which includes shocks, trends and seasonality, and on the other hand different social, institutional and organizational structures and processes that affect the way people gain access to these assets. With regard to the duality of structure (Giddens 1984), it is now important to notice that these structural conditions are determining factors for the self-employed activity. First of all, this structural framework is necessary for the business to be established and to run. Furthermore, it can have a limiting or enabling effect on the activity, which means, in the terminology of Klocker (2007), that it could function as an agency-thickener or an agency-thinner, e.g. by creating positive or negative trends in market prices, if a credit is granted by a bank or not. Finally, it will be assumed that through their activities these structural conditions can be changed to a certain extent, which is part of the examination.

In the given circumstances people develop livelihood strategies in order to achieve certain livelihood outcomes, e.g. to increase their financial capital or social networks. In this regard, a focus on *entrepreneurship* would be seen as a *livelihood strategy* that someone chooses in order to improve his livelihoods. A livelihood strategy often consists of combinations of activities, which Scoones calls a *livelihood portfolio* (Scoones 1998, 10). He identifies three different livelihood strategies:

- Agricultural intensification / extensification
- Livelihood diversification
- Migration

In applying these strategies to youth, some adjustments and expansions of these livelihood strategies have to be made. Krantz (2001) emphasizes with reference to Scoones' (1998) use of the term 'livelihood portfolios' the importance of unravelling the factors behind a decision regarding a certain livelihood combination. This strikes the core of the present evaluation, as the surveyed entrepreneurs can be distinguished by their different livelihood strategies and portfolios. Hence, these attributes are used as the main features for the type-building analysis. In this regard, it must be noted that this examination also considers the process of socialization of youth, which includes the accomplishment of different developmental tasks, e.g. by making educational and career decisions. Therefore, the question must be considered whether these young entrepreneurs plan their future in this activity or whether they want to escape and strive for a high prestige-job. Scoones (2017) points to the fact that youth have become a central policy concern with regard to the future of small-scale farming.

With respect to youth and the professions covered within the survey, the following livelihood strategies, which have been adjusted from Scoones (1998), seem appropriate (see also 6.6.2)[13]:

- Business growth: which is focussed only on the growth of the self-employed activity,

---

[13] These livelihood strategies are taken as the basis for the formulation of a hypothesis for the analysis (3.2).

which corresponds with Scoones' terminology of *agricultural intensification/ extensification*. In accordance with the analysis-design, the term *business growth* seemed more appropriate than *intensification* or *extensification*, as business growth is measured in terms of increased access to financial or physical resources.

- Business diversification: having a diversified livelihood portfolio, by either pursuing more than one (often unrelated) activity at the same time or by combining multiple agricultural activities, e.g. growing diversified crops and engaging in dairy farming. This strategy corresponds to *livelihood diversification*, but can also be pursued in connection with the business growth strategy.

- Education-escape: pursuing a higher educational degree in order to get a high-prestige job and to escape from the self-employed activity. This has been derived from the *migration strategy*. The term *escape* is used, as the surveyed entrepreneurs did not mention that they were planning to migrate – even though this might be inevitable for many of them – but considered quitting self-employed activity in order to pursue a high-prestige job.

- Business preservation: only keeping the status quo and not developing the business further. This strategy is added to the others in connection with Frese's (2000) elaboration on the reactive strategy (see 2.1.3).

An important question that occurs is whether to put the focus on the individual entrepreneur or the household level within the analysis. The sustainable livelihoods framework would be appropriate for both. Since the entrepreneurship training – which is the object of research – has been taken only by one person, the surveyed entrepreneur, and that person has been entrusted with responsibilities for the business, he or she will be the primary focus of the analysis. Nevertheless, in many cases the business was run as a family business. Therefore, when it comes to the enhancement of livelihood assets, the whole business – whether run individually or as a family – will be considered. But when it comes to future aspirations regarding education-al and career decisions, only the individual entrepreneur will be considered. However, the family will always be considered as a primary agent of socialization, being the immediate area of influence, as they are also part of the structural set-up and can function as agency-thickeners or -thinners.

Criticisms

A few criticisms regarding middle-range theories, which are also applicable to the concept of sustainable livelihood, have already been stated, such as the very schematic design which makes them inflexible with respect to the dynamic changes of environmental conditions, or the inability to comprehend the greater social context of problems (Müller-Mahn 2010, Dörfler et al. 2003). Considering this, it must be noted that for the structural design of the present analysis the sustainable livelihoods framework is only one building block. It takes into account the area of tension between agency and structure, but it lacks the ability to capture the duality of structure, how structural conditions change through the actions of individuals, and – vice versa – how individuals may adjust their actions to changing structural conditions. Krüger (2003) criticizes the fact that it fails to explain both motives behind actions and their causes if they fail to work out. This is an important aspect in considering youth in transition. Finally, this design does not capture the enhancement of capabilities – e.g. through entrepreneurship training – and how this impacts access to sustainable livelihoods. Therefore, it seems necessary to augment the sustainable livelihoods framework with further considerations regarding the empowerment of youth and their socialization. This will help to gain more insights into the personal motivations and aspirations of youth, and will also help to better understand the enabling and constraining effects of their immediate environment.

2.1.5 Youth empowerment and rural development

The main focus of this research project relates to the empowerment of disadvantaged youth through entrepreneurship training. At this point a closer look must be taken at youth empowerment and rural development. There are different concepts and means of understanding the key word *empowerment*, so it is necessary to elaborate which definition is most fitting for the present analysis. Thereafter, entrepreneurship training will be considered as an instrument to empower disadvantaged youth. This will now be performed within the context of Giddens' theory of structuration and the sustainable livelihoods framework.

2.1.5.1 The concept of empowerment

According to Navale and Sharma (2006, 67), empowerment has three components. Firstly, it is *multidimensional*, which refers to its relevance for different disciplines, e.g. sociology, economics, psychology, and social work. Secondly, it *occurs at various levels*, be it at the individual or the group/community level. Thirdly, it is a process, akin to a journey or a path. This broad understanding of empowerment underlines the need to put this term into context. The present analysis considers empowerment within the specific context of *youth empowerment* and the development of sustainable rural livelihoods. Therefore, the following definition given by the "International Fund for Agricultural Development" (IFAD 2005), which has also been adopted by a group of Indian experts (Roy and Saini 2009, 5)[14], who examined people's empowerment in considering sustainable rural development, will be discussed with regard to these three components. They define empowerment as "the ability of people, in particular the least privileged, (1) to have access to productive resources that enable them to increase their earnings and obtain goods and services they need, (2) and participate in the development process and the decisions that affect them." Roy and Saini further explain that one aspect without the other is not empowerment.

This definition certainly places empowerment within the context of the present study, as it focusses particularly on the *least privileged* and outlines their ability to have *access* to productive resources. This allows empowerment to link it with the sustainable livelihoods framework, which also considers *access* to different livelihood assets. This definition further specifies the target of the empowerment process, which emphasizes that empowerment enables the empowered individual to *increase earnings* and to obtain the necessary *goods and services*

---

[14] The volume of Roy and Saini (2009) is the outcome of a national seminar on "Empowering the Masses through Technical Application and Skills Training" held in December 2005 in Chandigarh, India. The participants were drawn from premier institutions such as the "National Institute of Technology" and the "National Institute of Technical Teachers Training and Research".

for it. This captures well what all active entrepreneurs included in the survey have explicated as their personal targets, which is to increase their income.

The *level* at which empowerment occurs within this definition is not the individual, but the collective, level. Navale and Sharma (2006, 74) point to the issue of "homogenizing all individuals sharing some particular characteristics" by making a social group or community the relevant unit of analysis. This concern has to be taken seriously, such that all kinds of personal criteria, such as occupational group, family background and education, will be considered within the analysis for each entrepreneur individually. Nevertheless, to have a group of people and not the individual as a reference point within this definition seems appropriate, as the target group is the youth sector from different socially deprived backgrounds, which share the common goal to attain advancement in life by the means of self-employed activity.

In further agreement with the definition, empowerment is understood as a *development process* or a *social process* (Verma 2006, 54). It involves an active part of the concerned actor in the participation in *the development process and the decisions that affect them*. Thus, it considers empowerment within the interaction of agency and structure. Garba (1999, 134) refers to two different viewpoints regarding the process of empowerment, namely the endogenous and the exogenous viewpoint. The first holds that only people could empower themselves. But there are facilitators from the outside supporting this process. The second view underlines the opportunity to bestow empowerment upon people from the outside. Taken together with Giddens' notion of power, empowerment was earlier defined as enabling the actor to exert power through the means of business activity, by drawing upon structural elements in the form of rules and resources (2.1.2). Power, as a structural element, is given from the outside. But it seems obvious that having the ability to exert that power involves the active engagement of the individual, and therefore the training provider is only seen as the facilitator of that empowerment process. The core assumption is that someone is not empowered per se simply by taking up an entrepreneurship training; rather it needs the active engagement of the individual. Therefore, it seems reasonable to assume that empowerment cannot simply be bestowed on someone from outside. This view is also supported by Jan (2009, 80), who states with regard to agricultural factors associated with female empowerment that "for women, where gender determines their access to resources and power, their empowerment begins when they not only recognize the systematic forces that oppress them but act to change the existing power relationship." Sarka and Sinha (2015, 106) also refer to the special difficulty faced by rural women to "leverage their empowerment to go beyond the confines of the village". Yet, they present a very specific case-study of two Indian women, who managed to do so by using the institution of a self-help group. One woman managed to enhance her dairy farming business, the other one her business in the area of textile design. The outcome was not only an increase of income, but also an increase of influence on political decisions, which affect them.

In the next section the definition provided by Roy and Saini (2009) will be further elaborated in connection with entrepreneurship training as the means to facilitate the process of empowerment.

2.1.5.2 Empowerment through entrepreneurship training?

Findings from research into youth transitions in developing countries show that "most young people must still accept that their educational and professional trajectories and biographical decisions in their way to adulthood, continue to be strongly determined by structural factors that, in most of the cases, are out of their own influence and control" (Bendit 2008, 362). In light of the previous observations on youth as agents, the question at hand is whether these limiting structural factors are really out of the influence and control of young people, or whether new perspectives would open up for them through entrepreneurship training to empower them to change structural conditions and to build up sustainable livelihoods, instead of continuing to be determined by structural factors.

At this point, the key insights regarding empowerment that have been gained so far will be replicated in order to apply them to the conceptual framework of this thesis. Empowerment is a social process, which cannot be bestowed on people from the outside, but it can be facilitated, in the present case by means of entrepreneurship training. In connection with the sustainable livelihoods framework, empowerment means to be able to gain access to sustainable livelihoods. Therefore, it means to productively combine the three elements which a sustainable livelihood is comprised of: capacities, assets and activities. Capacities can be directly increased through the training course, but they will have no value if they are not put into practice. This requires application to the various self-employed activities. Only then can the access to various livelihood assets be achieved.

Roy and Saini (2009, 5) further subdivide empowerment into economic, socio-cultural, political and self-empowerment. Regarding these different dimensions of the term, the issue of the terms of differentiation arises, as Roy and Saini (2009, 5) do not provide a precise definition of each dimension. For instance, when considering socio-cultural empowerment, they only refer to its necessity with respect to the "Indian socio-cultural context", especially in considering the empowerment of deprived castes and tribes, through which they gain social status. Yet, these categories largely coincide with the five livelihood assets provided by the sustainable livelihoods framework. According to the IFAD definition, one aspect of empowerment refers to the ability to have access to productive resources, which means access to livelihood assets. Therefore, these livelihood assets can be harmonized with the four dimensions of empowerment that run as follows:

- Economic empowerment includes the ability to gain access to financial capital, natural capital and physical capital.
- Socio-cultural empowerment includes the ability to gain access to social capital.
- Self-empowerment includes the ability to gain access to human capital.

Finally, political empowerment coincides with the second part of the definition of empowerment: the ability to participate in the development process and the decisions that affect them.

This process of empowerment is illustrated in Figure 2, which is an expansion of Figure 1. It includes considerations on how to gain access to sustainable livelihoods. The entrepreneurship training is considered as a medium to acquire knowledge and increase competences. This must be applied to self-employed activity. But the question arises what livelihood strategy the entrepreneurs have chosen. As a consequence, the self-employed activity has to be integrated in one of the following livelihood strategies (based on Scoones 1998), which are often a mixture of livelihood portfolios: e.g. sole focus on business optimization, business diversification (pursuing more than one self-employed activity at the same time), or educational escape (pursuing a higher educational degree in order to get a high-prestige job and to escape from self-employed activity). Now various activities are considered as routinized practises that take place within the interaction of agency and structure. The yellow arrows refer to the structure as being both the medium and outcome of practice. Empowerment is achieved if both criteria have been achieved, that is, the ability to gain access to productive livelihoods, which is visualized with a pentagon of five capital assets[15], and the ability to participate in the development process and relevant decisions, which requires an engagement in political, economic and cultural structures and processes. These are still to be discussed in further detail.

Figure 2: The process of empowerment in consideration with the sustainable livelihoods framework

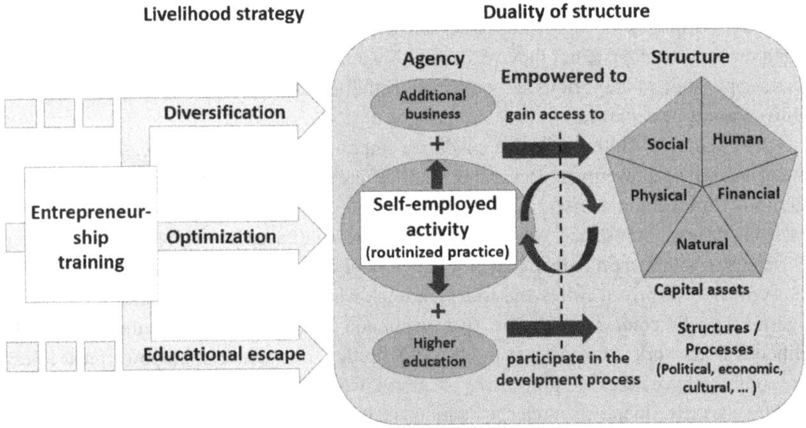

Source: Author's own draft, based on DFID (1999), Giddens (1984) and Scoones (1998).

2.2 Approaching youth from a pedagogical perspective

This sub-chapter approaches youth from a pedagogical perspective. Two aspects are here of the utmost importance: firstly, the socialization of youth, and secondly, the entrepreneurship training as an instrument for the empowerment of disadvantaged youth. Socialization is used as an umbrella term to denote education and learning processes (Stein 2017). The goal of the first part is to gain an understanding of both informal and formal education and learning processes, which the surveyed youth undergo in connection with different agents of socialization, such as the family, the school, and further higher educational or training institutions. In this respect,

[15] The pentagon has been taken from the sustainable livelihoods framework of the DFID (1999).

especial attention is given to such processes in relation to the various professions in which the surveyed youth received their entrepreneurship training. These socialization agents are also factors that determine the structural conditions, and therefore help to understand the duality of structure. But how do these socialization agents contribute to the "production and reproduction" of the entrepreneurial activity as social practice (Giddens 1984, 22)? An answer to this question, in turn, helps to understand the personal motivations, aspirations and perceptions of young entrepreneurs and to consider their transition into adulthood – including their school-to-work transition – within their cultural context.

Secondly, entrepreneurship training is looked at from a pedagogical perspective as an instrument for empowering disadvantaged youth, which includes an overview of the pedagogical approach of the different training providers that have been included in the survey. Lessons are hereby drawn from Gandhi's conception of basic education, as these basic ideas have been shown to have a normative influence on the developmental approach of present-day Indian social practitioners, such as the Drishtee Foundation.

Before the socialization of youth is properly considered, it seems appropriate to provide a definition of youth, which is not simply based on age, but which also takes account of the process of socialization that is undergone during youth.

2.2.1 Defining youth

The age-span of the entrepreneurs who are included in this research ranges from 18 to over 50 years old. The non-youth category has also been deliberately included as a comparison group. But where is one to draw the line between youth and non-youth? Currently, the National Youth Policy 2014 (MYAS 2014) defines Indian youth between the ages of 15 and 29. However, until 2013 the upper age limit was at 35 (GoI 2003). Since most of the entrepreneurs of this study participated in a self-employment training program under PMKVY[16] (MSDE 2016), which is targeted at giving skills to young people and for which the training locations had set an age limit between 18 and 35, it would make sense similarly to set the definition of youth up to the age of 35.

However, a definition which is solely determined by age falls too short of our research parameters. More important is to consider in addition certain developmental tasks that young people need to accomplish as they transition into adulthood. Klaus Hurrelmann (2012) identified four of these tasks specifically during the phase of youth, based on Havighurst's (1953) developmental task theory, which will be considered in more detail in sub-section 2.2.2.2 below. For now, only two aspects are pointed out which provide a basis for a definition of youth. Firstly, just as in Western countries such as Germany, the time spent in the educational system is continually increasing in India. This was notable also for the surveyed young entrepreneurs, in which seven out of 14 below the age of 35 had completed or pursued a Bachelor's degree at university. Secondly, unlike in Western cultures, marriage is not so much a choice as a cultural expectation, and would create social pressure if this developmental task was not accomplished by a certain age. Among the 14 surveyed young entrepreneurs, four had not yet completed their

---

[16] PMKVY – Pradhan Mantri Kaushal Vikas Yojana is an initiative of the Prime Minister of India for bringing skills to youth. For further details, see 2.2.3.4 below.

school or University education, and four were not yet married, which means that they were, so to speak, in a post-adolescent stage of life.[17]

Taken in consideration with the above, it makes sense to create terminologies for the following groups of people. For reasons of simplicity, anyone above the age of 35 is considered as "adult", and anyone between the ages of 18 and 35 is considered as "youth" or "young adult".[18] This group is next further subdivided. For the sake of distinction, those young adults who are not yet married and/or who have not finished their school or University education will be considered as "youth in transition". When the remaining group aged between 18 and 35 who did accomplish these developmental tasks are discussed, this will be specifically explicated in the text. These terminologies are illustrated in Table 1:

Table 1: Terminology for youth and non-youth sectors

Terminology		Age	Requirements
Adult		35+	-
Youth / Young adult	Youth in transition	18-35	Has not completed education and/or not yet married
	No specific term (it will be explicated in the text)	18-35	Has both completed education and is married

Source: Author's own draft.

2.2.2 Socialization of youth

In order to understand how to empower disadvantaged youth through entrepreneurship training it is helpful to take a detailed look at the socialization of youth. This is best achieved from a pedagogical perspective, where the interest is placed squarely on education and learning processes, although it is required to expand the focus and draw insights from psychological as well as sociological theories. At the core, socialization looks at the area of tension between the individual and society. Within the field researchers debate whether the individual personality is determined primarily by his or her genetic disposition or by the environment (Stein 2017). Modern interpreters favour a mediating position between both extremes, such as Klaus Hurrelmann, who defines socialization as "the process of the emergence, formation, and development of the human personality in interaction with the human organism on the one hand, and the social and ecological living conditions that exist at a given time within the historical development of a society on the other" (Hurrelmann, 1988, 2).

---

[17] The term ´post-adolescence` was first used by K. Keniston (1968) to describe a new and emergent life-phase between youth and adulthood in the third decade of life, which he identified among the rebellious youth of the 1960s, who behaved like adults in every respect but could not be described as such from an economic point of view. Zinnecker (1982) also used this term in the German debate.

[18] The terms "youth" and "young adult" are used interchangeably in order to refer to young people in the age group 18-35. The latter term seemed inappropriate for a specific description of youth who accomplished the developmental tasks such as qualifying and binding, as the "youth in transition" are also young adults.

For this research, attention is given above all to this second aspect, the impact of the social and ecological living conditions on the personality development of youth, although the impact of the human organism is not ignored. Socialization often takes plays in an unequal social setting, and therefore German researchers have recently given a lot of attention to the question of how social inequalities are produced and reproduced by socialization (see Bauer 2012, and the entire second edition of the *Journal for Sociology of Education and Socialization* 2007 on the subject, 'Socialization and Selection'). These aspects must be considered for the target group of this study, but not without paying due attention to the cultural conditions within the area of study (Trommsdorff 2008). This results in the question of how the youth sector handle different demands that are placed on them by their society, culture or economy, which in turn affects their personal aspirations as well as their motivations, attitudes and decisions towards entrepreneurial activity. Therefore, this sub-chapter begins with an understanding of youth as productive processors of reality, which draws on Hurrelmann's theory of the productive processing of reality. After this, linkage is made to cultural aspects and issues of inequalities and their connections with the developmental tasks that young people need to accomplish as they transition into adulthood, as well as the influence of socialization agents on the personality development of the youth subjects.

2.2.2.1 Processing reality in a productive way during youth

In their biographical transition into adulthood, young people are confronted with specific challenges, for which they need to activate their coping capacities (Hurrelmann/Quenzel 2016). Different expectations are placed on them, such as getting married or finishing education and attaining economic independence. Klaus Hurrelmann has created a meta-theory on the individual as a productive processor of reality, which looks at the tension between individuation, which means to develop one's own personality, and the individual's integration within the social structures of society. For this model he drew upon insights from both psychological and sociological theories, incorporating his key findings into ten propositions (Hurrelmann 1988; 2012). He further applied this model to the youth phase specifically (Hurrelmann and Quenzel 2016). The propositions regarding the developmental tasks and socialization agents have turned out to be highly relevant for this research and will be elaborated in more detail in sub-sections 2.2.2.2 and 2.2.2.3.

Hurrelmann's theory observes the processing and managing of an individual in light of the *internal* and *external* reality during an individual's life-span (Hurrelmann 1988, 45). The internal reality refers to the human organism, which includes the body and psyche and is substantially determined by one's genetic disposition. The external reality covers the social and material environment. The individual personality is the result of this process and is understood as "a person's particular organized structure of motives, attributes, traits, attitudes, and action competences" (Hurrelmann 1988, 45). The term *productive* refers to the fact that the individual always plays an active part in the management of his or her internal and external reality. Hurrelmann describes the individual as a creative constructor of personality with an ability for self-regulatory action (Hurrelmann 1986, 78; see also Hurrelmann 2012). This underlines the active dimension of the agent in accordance with Giddens' theory of structuration. The processing of reality is a dynamic form of activity, even though it often cannot be tacitly grasped, just as the production and reproduction of social practices, which are to a great extent part of the unconscious (Giddens 1984). Due to social disparities, the process of socialization is expected to take place differently among the youth included in this study, depending on their economic, social or educational status.[19] In order to understand this process of integration into society, a deeper look into the accomplishment of developmental tasks as well as the individual's various agents of socialization is necessary.

2.2.2.2 Developmental tasks of youth

In their transition into adulthood, the youth sector must manage to live up to different expectations that are placed on them. Based on Robert J. Havighurst's (1953) developmental task theory, Klaus Hurrelmann has defined four main developmental tasks which need to be resolved during the youth phase (Hurrelmann 2012, 79-80; Hurrelmann and Quenzel 2016, 24-28). The successful accomplishment of each of these tasks have both a personal and a societal dimension. On the one hand, each one supports the development of the individual personality, while on the other hand each supports the individual's ability to take autonomous actions, which ensures the individual's social integration by his or her taking over active membership roles in society. "Qualifying" includes the acquisition of competences, the completion of school, University or a vocational education and integration into the labour market. The successful accomplishment of this means to take over the active role of an economic citizen. It ensures economic reproduction both personally and for the whole society. "Binding" includes such activities as to find a spouse, to get married and to have children. If this task is accomplished successfully, it ensures biological reproduction on both individual and societal levels. "Consuming" includes the responsible handling of economics, leisure or media offered as a consumer and enables the individual to run a household. Finally, "participating" means to develop individual norms, values and convictions, which enable a person to participate as an active citizen in public affairs and to contribute to greater social cohesion.

For the surveyed young entrepreneurs in rural parts of Bihar, who are in the age group of 18-35 and who participated in an entrepreneurial activity, it can be presumed that all of them fulfil the requirements to accomplish each of these developmental tasks, since they have taken over active membership roles in society and are eligible for autonomous action. Therefore, it seems not necessary to have a deeper look at the tasks "consuming" and "participating". However, some of the young entrepreneurs are still in education and some are not yet married. Furthermore, the accomplishment of the tasks "qualifying" and "binding" has many implications for the personal motivations and aspirations of these young adults and the ways they engage in entrepreneurial activity. For these reasons, it is necessary to elaborate these two tasks in greater detail in connection with their Indian context.

Qualifying
In connection with this developmental task the question arises of what expectations young adults in rural parts of Bihar must meet. What influence do the families of young entrepreneurs

---

[19] This is supported by the findings of various researchers, who have pointed to the fact that socialization processes contribute to the recreation of social disparities (cf. e.g. Bauer 2012; Grundmann 1994; Grundmann and Hoffmeister 2007; Hurrelmann et al. 2008).

have on their academic or career-related choices and their engagement in entrepreneurial activity? A youth survey in India has revealed that more than 40% of both male or female Indian youth value either strong or very strong parental authority in their decision-making. The majority of them claim that their parents would take all educational or career decisions for them (DeSouza et al. 2009). Moreover, those who prefer limited parental authority would still consult their parents in their academic or career choices. It seems reasonable to assume that the percentage of youth who value high parental authority is even higher among those growing up in traditional villages in Bihar.

As for entrepreneurial activity, it must be noted that most of the surveyed young entrepreneurs in rural parts of Bihar grew up in a home with a traditional family business, so eventually they became involved in it as well. For instance, all of those who had taken a training in dairy farming came from traditional dairy farmers' homes. The same is true for handloom workers. Often, the entire village was part of a cluster for that specific occupation. For these young adults, it was more a "natural" process than a voluntary decision to become engaged in entrepreneurial activity, as they were introduced to it very early on as part of an informal learning setting within the family, which could be classified as "functional education" (see Kern 2017, and further 2.2.2.3 below). Furthermore, in taking into consideration the fact that in India such traditional activities are usually linked with the caste group, and are therefore often characteristic of a lower social status, the question arises whether the early participation in such family businesses prevents them from a good school education and more open career options[20].

Saraswathi (1999, 213) observes an "adult-child continuity" among children growing up in the lower or middle social class in India, as children take on adult roles very early on. She notes that, in rural areas, boys are very often compelled to help to contribute to the family income by the time they are 13 or 14 years of age, whereas girls are expected to become more involved in housework and take care of younger siblings by the age of eight or nine. This also often takes place at the expense of their education. For this reason, such children or teenagers also have little opportunity to engage in leisure activities, which leads to the conclusion that "when there is an absence of segregation of youth and adults, because education is through apprenticeship for participating in adult activities, socialization into work is a continuous process, such as in artisan families or agricultural labourers, and leisure time is programed by the entire village community, then the specification of an adolescent stage has little meaning" (Saraswathi 1999, 223). This observation supports the assumption that, also in rural parts of Bihar, most young entrepreneurs from the lower or middle social class did not have the privilege to experience a distinctive youth phase, let alone a post-adolescent phase of life, if they extended their stay in the educational system up to the third decade of life (see 2.2.1 and Keniston 1968; Zinnecker 1982).

This leads to the important observation that in India great educational achievements have been attained over recent decades, which did not leave out Bihar, where an increasing

---

[20] The Indian caste system, which evolved over centuries, is based on a division of labour. Narrowed to the main castes, the highest caste, the "Brahmins" are the priestly class, followed by the "Kshatriyas" (the warrior class), the "Vaishyas" (the merchant and peasant class) and the "Shudras" (labouring class). At the bottom of society are the "Untouchables" who perform jobs that are traditionally perceived as unclean. Although the caste system has officially been rescinded, it still stipulates the lives of people particularly in rural areas in large measure (Deshpande 2014).

number of young men and woman from lower social classes have obtained the chance not only to complete their secondary school education, but also to pursue a University degree. DeSouza et al. 2009 found that the vast majority of the Indian youth sector strives for well-paid and secure government jobs or other high-prestige jobs. Among the surveyed young entrepreneurs two out of ten from below the poverty-line had already completed a Bachelor's degree. This should unquestionably open a whole new range of professional opportunities for them. And yet, they were still stuck in relatively unprofitable self-employment activities, which indicates that they remained bound to the traditional village structures. This leads to the research question of how the entrepreneurial activity fits with their academic or career aspirations. Moreover, the question arises whether socially deprived young adults could profit from a meritocratic principle, according to which their educational achievement would give them a chance to overcome confining traditions and climb the social ladder (Bauer 2012).

Binding

When it comes to marriage it seems necessary to take into account certain differences between "Western" and "Indian" cultures. In Western cultures the average age of marriage has become increasingly higher, whereas in India – especially in rural parts, as in the target regions within Bihar – both men and women often get married at a very young age, normally following the Hindu tradition of an arranged marriage. Among the interviewees, the youngest married person was within the age group 20-25. Unlike in Western cultures, the married couple usually move to one of the parents' homes. Saraswathi notes with regard to dating, marriage and the duration of adolescence, that "marriage is clearly a family affair, a union of families and not a union of only two individuals" (Saraswathi 1999, 229). Therefore, from a sociological viewpoint, "adolescence is a period of transition from the family of orientation to the family of procreation" (Saraswathi 1999, 223).

Saraswathi (1999, 230) further finds that in contrast to Western cultures young people, including college students, prefer to "let parental wisdom prevail in this matter of marriage". DeSouza (2009) confirms this opinion on marriage and finds that 65% of all Indian youth let the final marriage decision be taken by the parents. In addition, it is customary to arrange marriages within the same social class, with the result that marriage functions as an institution that contributes to the reproduction of social inequalities. According to DeSouza et al. (2009) 67% of Indian youth consider it important that marriage should take place only within one caste.

For both developmental tasks qualifying and binding, it can be concluded that in spite of visible trends of individualization, which become obvious in an increasing number of young adults who do not accomplish these developmental tasks consistently with traditional norms, the hypothesis still appears to be true that they cannot free themselves from traditional contexts of norms, values, and classes (Beck 1986).[21] Therefore, it may appear that they are – without

---

[21] The German sociologist Beck (1986) postulated the individualization paradigm, according to which individual life opportunities are no longer destined by the social class, due to the modernization of welfare and new forms of employment in Western societies after the Second World War (principle of structure), but instead individual biographies are increasingly structured by individual delineations of action (principle of action). For the given context of disadvantaged Indian youth from rural backgrounds, it appears that the first assumption cannot simply be transferred, as they cannot free themselves from traditional norms and classes. At the same time,

being married or having completed their school or University education – in a kind of post-adolescent life-stage in which they experience a "psycho-social moratorium" (Erikson 1968/1974). But their living reality suggests that they did not even have the privilege of experiencing a distinct adolescent life stage, given their early engagement in adult responsibilities, such as participation in family businesses (Saraswathi 1999).

In light of these considerations, the time when the developmental task "qualifying" has been successfully accomplished by a young entrepreneur becomes fluent if and when the entrepreneurial activity is completed as a part of a family business, in which children are often expected to engage from an early stage. In many cases the economic reproduction of the family is not ensured with one, but several, entrepreneurial activities. For the sake of identifying the "youth in transition" (see 2.2.1) within this survey, the accomplishment of this developmental task is measured by the completion of the school or University education. The accomplishment of this developmental task "binding" is conversely measured by the marital status of the entrepreneur.

2.2.2.3 Socialization agents

In line with Giddens' theory of structuration, this section provides a detailed look at socialization agents as part of the structure. As youth are considered between agency and structure, the present focus here lies not on how youth can change structural conditions, but on the existing structure into which they are born, and with which they become acquainted during their primary and secondary socializations. The first process of socialization is to be brought up at home, and the second has been described by Berger and Luckman (1967, 158) as the "acquisition of role-specific knowledge", which consequently includes vocational education and training or University education. From a pedagogical perspective both the first and secondary processes of socialization are of interest in order to comprehend both formal and informal processes of education and learning.

The socialization agents on which young adults depend during their process of personality development, being based on Urie Bronfenbrenners' (1981) theory of ecological development, consist of three layers (Hurrelmann 2012). The family is the most influential factor and is therefore considered as a primary socialization agent. Secondary socialization agents comprise all educational institutions, which include day-care centres, schools, any type of vocational education, and training facilities and universities. Further systems that are relevant for the socialization process are ranked as tertiary socialization agents, such as the occupational sector, intimate partnerships, one's circle of friends, media and further political or religious institutions. There are small differences in the perception of this third category. For instance, Kern (2017) ranks in her broad overview only schools among secondary socialization agents and Universities among tertiary ones. For the present work, however, it is not important to make a clear division between all kinds of primary, secondary and tertiary socialization agents, but to identify the most relevant ones and to understand how they affect the process of socialization,

---

they do have greater educational opportunities, and thus seem to have greater opportunities for individual delineations of action (Beck 1986, 119; Bauer 2012, 78).

the individual's engagement in entrepreneurial activities, and the career decisions of rural Indian youth. Closer attention will be given to the following agents in light of their specific cultural and rural Indian context.

The family as primary socialization agent

The family in its Indian context comprises not simply the nuclear family, but all members living under the same roof, which usually includes further relatives. Looking at the family as part of the structure, it must be highlighted that this institution can have both limiting or enabling effects on youth personality development. The different dimensions of personality development on which the family has great influence are knowledge, motives, feelings, needs and competences (Hurrelmann 2012, 66). Hurrelmann presumes that the better the economic condition of a family and the educational degree of the parents, the "richer" is the socialization process within the family. The strong influence of the family on youth has already been mentioned in connection with the accomplishment of the developmental tasks "binding" and "qualification". Saraswathi (1999) points out limiting factors in considering the upbringing of boys and girls in families from poorer sections of Indian society. Girls are considered as "guests" within their maternal home, for they will leave the home by the time they are married (Saraswathi 1999, 215). They have to engage to a great extent in the house-keeping as well as in sibling care, and are expected to perform these tasks with great proficiency and modesty. Boys are often dictated by the "compelling need to earn, before one can learn" (Saraswathi 1999, 215). These aspects provide important insights regarding the structural conditions of disadvantaged youth, even if they have not been explicitly expressed in the course of the evaluation.

From a pedagogical perspective, a major interest is placed not on the educational task of nurturing and parenting the children, but on the informal processes of learning or functional education, which prepare children to take up certain professions. This is evident, for instance, in the case of handloom working and dairy farming, to which the adolescents are introduced at an early stage (Stein 2017). In order to comprehend these processes, it makes sense to consider the family in connection with the village community. Even though the village community would be considered as a tertiary socialization agent, the boundary between the family and village community is often more fluid in rural Indian settings, due to the strong family relationships within the village, their being of the same caste, and their often having the same traditional professions. Therefore, individuals have a high identification with their village community. Often, the village is additionally part of a certain business cluster, e.g. a handloom cluster or dairy farming cluster, which has a determining effect on young people's engagement in such activities. In order to grasp the family as a primary socialization agent and the greater village environment within the survey, all the entrepreneurs have been visited within their village. The interviews usually took place at their homes in the presence of a parent.

Training providers as secondary socialization agents

The school is undoubtedly among the most important secondary agents of socialization, yet it is not the focus of the present examination, and therefore it will not be discussed at this point.

Following Hurrelmann's differentiation, training providers will be considered as secondary so-cialization agents, as they represent an educational institution. In viewing them as part of the structure, it must be noted that all the surveyed training providers were closely connected with the village community, and as such they aligned their activities with traditional village profes-sions. The training was, moreover, held in various village locations to ensure proximity to the village people. In the case of Drishtee, training is only one part of a further reaching develop-ment strategy, which aims at developing sustainable communities. Therefore, training providers often function not only as facilitators of entrepreneurship training, but also play a vital role in the empowerment process of the target group and contribute to the improvement of structural conditions. For the present research, the question arises to what extent each of the training pro-viders included in this survey fulfil this role. One important aspect to consider is the handhold-ing between the training provider and their former trainees. Each of the training providers will be looked at in greater detail from a pedagogical perspective in the sub-section 2.2.3.

In considering processes of education and learning, training providers obviously have a function to impart learning within a formal setting, a process which occurs in the second phase of socialization. As already outlined in consideration with Giddens' theory of structuration, a part of the examination is focussed on the increase in discursive consciousness through the training, and how this affects the individual's capability to cope with structural conditions, and thereby fosters the process of empowerment.

Tertiary socialization agents

There are further tertiary socialization agents that seem worthwhile to consider in this connec-tion. At this point they have only been mentioned in passing, but will be addressed in greater detail in due time. These agents are:

- community-based organizations, which include farmer clubs (these will be considered under 2.3.2);
- banks (which have the important function of providing credits to small-scale farmers and entrepreneurs, and will be considered in the context of community-based organiza-tions);
- the economic system (which will be considered for each profession separately under 2.3.3);
- government institutions, such as Panchayati Raj institutions (public bodies of local self-government); and
- Universities.

The following image (Figure 3) illustrates the different primary, secondary and tertiary social-ization agents and their influence on rural entrepreneurs in India. The primary and secondary socialization agents have already been considered in detail. Tertiary socialization agents, such as the economic system, Universities or the government can have a supportive effect on the entrepreneurial activity, for instance if the government promotes such activities through entre-preneurship training schemes, if a University degree is chosen in an academic subject that can be applied to the business activity, such as agricultural sciences, or if the economic system is

33

used to increase the business. These agents can also have a pull-effect, drawing young people away from a self-employed activity, for instance if they pursue a higher educational degree in order to obtain a government job, or a high-prestige job, for which they would need to migrate into the city.

Figure 3: Socialization agents and their influence on rural entrepreneurs

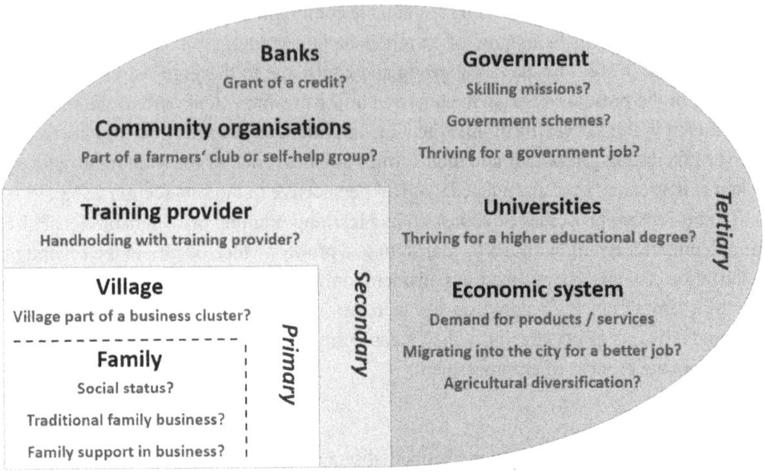

Banks
Grant of a credit?

Government
Skilling missions?

Community organisations
Part of a farmers' club or self-help group?

Government schemes?
Thriving for a government job?

Training provider
Handholding with training provider?

Universities
Thriving for a higher educational degree?

Tertiary

Village
Village part of a business cluster?

Economic system
Demand for products / services

Family
Social status?

Migrating into the city for a better job?

Agricultural diversification?

Traditional family business?

Family support in business?

Primary

Secondary

Source: Author's own draft, adjusted from Hurrelmann (2012).

2.2.3 Entrepreneurship training in India: an opportunity for disadvantaged youth?

Thus far entrepreneurship training has been examined from a pedagogical point of view. Now, the focus is shifted to the Indian context in general, and specifically to the context of rural parts of Bihar. At first, it seems necessary to establish clarity on how the terms *entrepreneurship* and *self-employment* are used within the present analysis. In order to understand the development approaches of present-day providers of entrepreneurship training in India, it is helpful to take a glimpse back at Gandhi's conception of basic education, which integrated the teaching of hand-icrafts into the school education. In this Gandhi envisaged reviving the village industry and revolutionizing the Indian education system. This will be related to current initiatives of entre-preneurship training within the context of India's skill development mission. Finally, the se-lected providers of entrepreneurship training and their pedagogical approaches to empower dis-advantaged youth, as well as the respective training curricula, will be studied.

2.2.3.1 The issue of defining self-employment and entrepreneurship

Due to a great number of theories, definitions, and conceptions with regard to the term *entre-preneurship,* it is difficult to differentiate the two terms, *self-employment* and *entrepreneurship,* from each other (Parker 2004)[22]. Before the potential of entrepreneurship training in light of India's skill development mission can be considered, it must be made clear what these terms in our Indian context actually mean. Therefore, in what follows they will be examined both from a more general theoretical perspective and within the specific context of the present analysis.

To begin with, there is a broadly agreed understanding of self-employed individuals as "individuals who earn no wage or salary but who derive their income by exercising their profession or business on their own account and at their own risk" (Parker 2004, 6). These criteria are undoubtedly met in relation to the examined self-employed activities. Yet, the "National Commission of Enterprises in the Unorganised Sector", under the chairmanship of Dr. Sengupta, explains in their report that "self-employment is a 'catch all' category and can be broadly divided into those with physical capital and human capital and those without. The lower levels of self-employed are the working poor, often closer to casual workers, in such occupations as street vending, rickshaw pulling, beedi rolling[23], etc." (Sengupta 2007, 18). In rural parts of Bihar, a majority of 50% is considered self-employed (in all of rural India, it is 57%), 48% are casual workers (in all of rural India, 35%), and only 5% of the workers fall under the category "regular wage/salaried employee", which is far below the respective figure of 16% for all of rural India (NSSO 2011). This sobering reality of extreme wide-spread poverty in rural parts of Bihar, in spite of the fact that a majority of the workers is involved in self-employment-activities, shows that being "self-employed" does not automatically mean to be "empowered". Still less does it automatically mean that being "self-employed" means to actively change the structural conditions through one's action. Thus, the question can be asked: can the surveyed individuals, who can certainly be considered as self-employed, also be considered as entrepreneurs?

In its introduction to the course modules for the Rural Self-employment Training Institutes (RSETI, see 2.2.3.5), the National Institute of Rural Development (2011) makes an important differentiation between income-generating activities, self-employment, and entrepreneurship. An income-generating activity could refer to any money-making activity. In contrast, a self-employed activity describes the second stage, which is attained if "a person fully utilizes his or her entire time and energy in carrying out the activity. The income generated is on a continuous basis and the activity has a definite shape" (NIRD 2011, 21). Entrepreneurship, in turn, describes a next stage: "Entrepreneurship is the character, practice and/or skill of an entrepreneur. An entrepreneur is a person who organizes, manages and assumes the risk of a business. Accordingly, entrepreneurship refers to identifying/innovating ideas, product and services; mobilizing resources; organizing production/service and finally, marketing them with constant strive for growth and excellence" (NIRD 2011, 20). Thus, the main difference between self-employment and entrepreneurship lies in the utilization of certain entrepreneurial qualities, which exceed the objective of solely earning money.[24] These characteristics find support from

---

[22] For an extensive overview of different approaches in research on entrepreneurship, see for instance Bruyat and Julien 2001, Parker 2004 or Low and McMillan 2007.

[23] "Beedi" is an Indian type of cigarette.

[24] NIRD (2011) identifies 15 entrepreneurial competencies, which will be looked at in consideration with the curricula of the training providers included within the present analysis under 2.2.3.7 below.

Miller (1983, 771), who identifies an "entrepreneurial firm" by its "product-market innovation", by its will to undertake "risky ventures", and by its "coming up with proactive innovations". Yet, with regard to the question of what distinguishes entrepreneurship from self-employment, further insights can be gained from Carland et al. (2007), who distinguish entrepreneurs from small business owners. Relying on the classical concepts of innovation and entrepreneurship of the economist Joseph Schumpeter (1934), Carland et al. conclude that innovation is the most distinguishable factor. They present a definition of both these key-terms. On the one hand, they highlight the fact that a *small business owner* is "an individual who establishes and manages a business for the principal purpose of furthering personal goals". In contrast, they define the *entrepreneur* as "an individual who establishes and manages a business for the principal purpose of profit and growth. The entrepreneur is characterized principally by innovative behaviour and will employ strategic management practices in the business" (Carland et al. 2007, 79). This conception resonates with a distinction made by the Global Entrepreneurship Monitor (GEM) between "necessity entrepreneurs" and "opportunity entrepreneurs" (Reynolds et al. 2002). Opportunity entrepreneurs are characterized as improvement-driven and geared at exploring new market opportunities, whilst necessity entrepreneurs are characterized – like the small business owners (Carland et al. 2007) – as driven by personal needs, e.g. unemployment or unsatisfactory employment situations.

In applying these findings to the context of young entrepreneurs in rural Bihar, the category of the "small business owners" would certainly fit with many of the surveyed entrepreneurs. Yet, it seems too simplistic to characterize them monolithically as being solely self-employed individuals, but not entrepreneurs.[25] This perception is contradicted by the fact that the surveyed "small business owners" were also striving for profit and growth.[26] Accordingly, they also cannot be simply marked as "necessity entrepreneurs". Williams and Williams (2011) also make the criticism that this dichotomy between necessity and opportunity entrepreneurship would not appropriately reflect the entrepreneur's motives, which may change over time. In particular, this concern cannot be ignored with regard to young entrepreneurs who may not yet have accomplished the developmental tasks of qualifying and binding. Hence, it is likely that the surveyed young entrepreneurs who acquired entrepreneurship training – even if they may have been driven by personal necessities in the first place – will develop entrepreneurial competencies and proactively utilize new business opportunities. On the other hand, if they fail to come up with "innovative behaviour" and "strategic management practices"[27], it rather seems to be an issue of empowerment. This underlines the need for entrepreneurship training, which is targeted at developing certain entrepreneurial competencies.

Thus, it seems more reasonable to pay attention to Miller's (1983) notion of proactiveness, which ties together the previous insights into the importance of the development of personal initiative in connection with gaining access to sustainable livelihoods. Lumpkin and Dess (1996, 146) understand proactiveness as "taking initiative by anticipating and pursuing new

---

[25] However, the identified livelihood strategies (see 2.1.4.2), which have been derived from Scoones (1998), may allow the conclusion that those entrepreneurs following a preservation strategy tend to be "self-employed" to a greater degree, while those following a "business growth strategy" tend to be the true "entrepreneurs".
[26] This will be shown throughout the analysis in Chapter 6.
[27] See Carland et al. (2007) above.

opportunities by participating in emerging markets". Personal initiative certainly includes innovative behaviour, risk-taking and innovation, and thus can be taken as an identifier for entrepreneurial competencies and behaviour, allowing us to make a distinction between reactive and proactive entrepreneurs. Yet, for the purposes of the present analysis, it seems appropriate to adopt the concept of personal initiative from Frese (2009), who did not apply proactiveness exclusively to new business innovations, but understood it from a psychological point of view in a broader sense as a means of overcoming barriers and constraints of any kind. These would, naturally, include any proactive entrepreneurial behaviour, such as applying training inputs, engaging in social networks, availing financial support, and the like (see 2.1.3).

With regard to the usage of the terms *entrepreneurship training* and *entrepreneur/entrepreneurship* on the one hand, and *self-employed/self-employment* on the other hand, these terms will be used interchangeably within the present dissertation, now that the key issues of defining and distinguishing them has been completed. Yet an important remark must be made regarding the usage of the term *young entrepreneur*. Based on a distinction by GEM between *young entrepreneurs* and *nascent entrepreneurs*, GEM understands nascent entrepreneurs as being in the process of starting a new business venture, and *young entrepreneurs* as individuals between the age of 18-64 who have started a business not more than three and a half years from the time of study (Sternberg and Bergmann 2003, 11). Within the present analysis, this understanding of young entrepreneurs will not be applied. Whenever the term "young entrepreneur" is used, the word "young" will always refer to the entrepreneurs' "young" age between 18-35, but not to the "young" age of the respective business undertaking. This makes sense in light of my dissertation's specific focus on entrepreneurship among "youth".

2.2.3.2 Drawing lessons from Gandhi's conception of basic education

Gandhi's pedagogical ideas belong to an academic debate that continues until the present day (Lang-Wojtasik 1997/2002; Dasgupta et al. 2010). An understanding of Ghandi's conception of basic education is necessary, because the underlying ideas and convictions still attract present-day providers of entrepreneurship training and influence their approaches regarding youth empowerment. Gandhi's objective was not only to free Indian schools from the British education system, but to establish a holistic theory of pedagogy. This includes vocational education as much as theoretical classes, and thus makes it relevant for the present work. By placing entrepreneurship training for disadvantaged youth in focus, this sub-section aims to carve out how Gandhi's pedagogical ideas provide a perspective for empowering disadvantaged youth, and to what extent the agencies of various present-day Indian agents concerned with entrepreneurship training are influenced by them.

In order to grasp Gandhi's pedagogical ideas, it is important to understand how he envisioned independent India.[28] His struggle to free India from British colonial rule was closely linked with the concept of *swaraj*, which can be translated as "self-rule" or "self-government". Gandhi envisioned a self-reliant and self-governing India with a decentralized government and economy, in which villages fulfil a vital function (Babu 2011). This vision could be summarized in the notion of "village *swaraj*", on the idea that "India lives in her villages and not in her

---

[28] Gandhi refers to an India encompassing the area of the present states of Bangladesh, India and Pakistan.

towns".[29] According to this philosophy, the 700 000 Indian villages must also be self-governed by village panchayats (governments of five) and self-sufficient with respect to basic needs like food and cloth. Thereby, Gandhi propagated the small-scale cottage industry as the essential industry, which had to be revived in every village. But he also states that "*swaraj* is to be obtained by educating the masses to a sense of their capacity to regulate and control authority".[30] This statement underlines the strategic importance of Gandhi's concept of basic education – which he called *Nai Talim* (New Education) – for the achievement of *swaraj*. This educational concept will be introduced in more detail in the following paragraphs.

Gandhi developed the concept of basic education in contrast to the British education system (Molt 1970), which he criticized as being solely oriented at the education of the intellect. This would solely stimulate self-realization through career-orientation and lack any greater conception of character-building. It would result in the economic promotion of one world-view, but leave out the masses. Furthermore, the British education system, which was based on a foreign culture, lacked any reference to the Indian culture. In addition, teaching was in English, which – according to Gandhi – was a sign of slavery.

Instead, Gandhi advocated the education of both the heart and hand by advocating a vital connection between education, national culture and religion (Molt 1970). In this regard, he said that it is most effective if it takes place in harmony between the home of the parents and school. The task of education is not limited to the children in school, but includes the entire village (Molt 1970). Gandhi describes this holistic educational approach as "an all-round drawing out of the best in child and man — body, mind and spirit".[31] The two key elements he posited were thus character building and vocational education. The first served as justification for a literary education, as its outcome must be a sound character. The latter has the same importance as theoretical teaching. Gandhi explains that "literacy is not the end of education nor even the beginning. It is only one of the means whereby man and woman can be educated. Literacy in itself is no education. I would therefore begin the child education by teaching it a useful handicraft and enabling it to produce from the moment it begins its training".[32]

Gandhi used a practical teaching method based on *correlation* (Molt 1970, 49), which places handwork at the centre of teaching and other subjects which have some interdisciplinary connection with it.[33] In this Gandhi used the spinning wheel (or charkha) in school as a symbol of service to the nation, and connected it with a special social and spiritual meaning. Other subjects, like numeracy, history or geography could also be learnt from practical teaching, for instance the exercise of how to spin.[34] This integration of vocational education into the curriculum prepares children to remain in the tradition of the village. Gandhi did not reject the caste system exactly[35] but encouraged the children to learn the profession of their parents,

---

[29] In: The Hindu, March 19, 1925, *The Collected Works of Mahatma Gandhi* (CWMG), Vol. 30, 412.
[30] In: Young India, January 29, 1925, p. 40-41, CWMG, Vol. 30, 159.
[31] In: Harijan, July 31, 1937, CWMG, Vol. 72, 79.
[32] In: Harijan, July 31, 1937, CWMG, Vol. 72, 79.
[33] In Harijan, October 16, 1937, Gandhi 1951, 59.
[34] Other handicrafts have also been considered, such as gardening. According to Gandhi the *charkha* was most appropriate (Speech at Education Ministers Conference, July 29, 1946. In: Harijan, August 25, 1946, CWMG, 91, 377).
[35] Gandhi believed in *Varna Dharma* (caste system), through which authority he demanded the elimination of untouchability (see the 13 point program: In Harijan Sewak, September 14, 1940; CWMG, Vol. 72, 451).

become self-reliant, start their own family and integrate themselves into the village. In this way, the new school curricula served the villages. Thereby the school had economical autarky (Molt 1970).

Gandhi tested his ideas in practice by establishing different *ashrams*, namely the Phoenix Farm (established 1904) and Tolstoy Farm (1910) in South Africa, and the Sebarmati Ashram (1915) and Sevagram Ashram (1936) in India. An *ashram* is a religious community and a model for a village, where the responsibility of education lies in the community. An ashram comprises of houses in which families live. Attached to the community are production centres and the school. The curriculum within the ashram school was divided into three phases (Molt 1970). Up to the age of eight the school followed a co-educational model, in which both boys and girls were taught together. At this stage, handwork was introduced appropriate to the age of the pupils. In the age-group of nine to 16 the co-educational model continued, and, besides one's mother-tongue, Sanskrit was taught for Hindu children and Arabic for Muslim children. At this stage children were encouraged to learn the profession of their parents. Other general skills were taught, such as general education, cooking or sewing. In the last stage, for pupils aged from 16 to 25, the youth received individualized education. But Gandhi emphasized that the real education starts when the child leaves the school, as experience is the best teacher.[36]

Even though Gandhi's conception of basic education, which he introduced at the Wardha conference in 1937, was widely accepted by the Congress[37], the realization of it after 1947 in the newly independent India was disappointing and deficient. This section aims not to track the history of basic education or to discuss the reasons for the lack of acceptance (for this see for instance Prakasha 1985), but to understand its relevance for the present work. However, one important remark in passing should be made. Instead of following Gandhi's *Naim Talim* as mainstream education, the model of general education was followed on the basis of the acquirement of knowledge, which led to a devaluation of vocational education on the grounds that general education was not fit for the poorer segments of society (Palanithurai 2016).[38]

A parallel between Gandhi's new school and entrepreneurship training nowadays is that both prepare the taught subject for self-sufficiency. The school was also a small business, in which the products made by the children were sold. But the school was only meant to meet basic needs, not to make a lot of money, which would contradict Gandhi's conviction. Another important aspect is the function of the school to purposefully prepare the children to follow the village tradition by learning the parents' profession.

Taken in connection with current trends of entrepreneurship training, the question arises whether the training should be still focussed on preparing youth to stay in the village or not. Molt explains the failure of Gandhi`s conception of basic education not as a shortcoming in his

---

Competition and exploitation had to be reduced to a minimum, and cooperation and social consciousness had to be enhanced (Molt 1970, 64).

[36] Lang-Wojtasik (2002, 192) and Muniandi (1985, 11ff) present Gandhi's idea of lifelong learning as adhering to the following stages: Pre-Education (age 3-5); Basic Education (6-14); Post-Basic Education (15-18); Adult Education (Social Education) and the Rural University.

[37] For details see the Speech at the Wardha-conference (in: Harijan, October 30, 1937; CWMG, 66, 273). The resolution encompassed inter alia compulsory education up to Grade seven, teaching in the mother tongue, and the learning of a handicraft at the core of the educational agenda.

[38] This perception is also reflected in personal interviews conducted for the present work with entrepreneurs, trainees and training coordinators in Bihar.

pedagogical theory, but by the fact that "the social dynamic moved into a different direction", which brings him to the conclusion that "if basic education wants to be pedagogically effective, it must be socially realistic" (Molt 1970, 76, my own translation). The question of social relevance is appropriate, as the adherence to traditional customs stands against trends of globalization and the thriving of young people to attain higher educational degrees which prepare them for a professional career. But irrespective of these trends that may cause many young people from rural parts of Bihar to migrate to the big cities, a large share of them remains in villages and within the traditions of the family, especially young people from poorer sections of the society. For them self-employed activities must present a realistic perspective by which to build up sustainable livelihoods. Therefore, it seems relevant to consider entrepreneurship training within the context of basic education under different aspects, such as the choice of subjects in which these courses are offered, how these professions are linked to the village community, and what effects the entrepreneurship training has on both the empowerment of the individual and the development of the village community through revitalization of the profession. This will be subject to further elaboration in 2.3.2 and 2.3.3 below. Since the present work does not focus on school education, the principle of correlation, or how other subjects could be taught through the learning of handicraft, is not part of this elaboration.

It is self-explanatory that the *charka*, which for Gandhi was a symbol of the independence movement, does not hold the same importance anymore, and therefore will not be considered in any particular way, even though charka workers have been included in the present survey. In a similar study, Vinoda Bhave applied Gandhi's pedagogical ideas but used agriculture as the practical subject instead of spinning (Molt 1970, 62). Likewise, the present research deliberately takes its cue from Gandhi's convictions on the opportunity of empowerment through entrepreneurship training in different village-related – agricultural as well as non-agricultural – subjects, with the aim of exploring its efficiency for improvement of sustainable livelihoods and the handling of structural limitations.

2.2.3.3 Entrepreneurship training and India's skill development mission
In an allusion to Gandhi's vision of autarchic villages, this sub-section tries to carve out why a focus on entrepreneurship training in agricultural subjects and subjects related to traditional village industries are relevant for India's skill development mission today. A look at the demographic set-up of India underlines the urgency of empowering its young generation through education and training and of generating new job opportunities. India has the world's largest youth population, which presents a demographic dividend due to a declining dependency ratio up to the year 2040 (according to the United Nations Population Division Projections, Mitra and Verick 2013; Planning Commission 2008/2013; Mitra and Nagarajan 2005; Navaneetham 2010). Given the very young population in Bihar, which has a median age of 20 years – as opposed to 25 years for all of India (Census of India 2011) – the need is even greater to harness the demographic dividend through qualification initiatives in very rural regions such as Bihar. Otherwise, as Sankaranarayanan said, there would be "create[d] an army of unemployed youth in the country, leading to a host of socioeconomic problems with wider consequences" (2011, 56).

India's skill development mission

In order to face these challenges, the Indian government defined the ambitious goal of qualifying half a billion people by the year 2022 (National Policy on Skill Development, MoLE 2009). The goal of achieving "education for all" has been expanded from this aim to the secondary stage. But how is India to accomplish this mammoth task in the area of skill development in a state like Bihar, which is dominated by agriculture and marked by an absence of industry, and is consequently lacking in job opportunities in the non-agricultural sector? In addition, the agricultural productivity in Bihar is one of the lowest in India. 93% of the farmers are small and marginal, which is a higher percentage than for all of India (84%). Only 29% of the households own any land and the average landholding size is only one acre (NCEUS 2008). Therefore, in rural parts of Bihar people strive hard to access sources of livelihoods, which is reflected in a high work participation rate. This often keeps young people from completing their school education or acquiring vocational skills (Mitra and Verick 2013). As a consequence, Bihar is the Indian state which contains the highest rate of people who migrate to other parts of the country (National Sample Survey 2007–2008; see Bhagat 2016). This concerns especially young people from rural parts of Bihar, who find no future prospects for themselves in the agricultural sector (Kumar and Bhagat 2016; Rigg 2006). This brings the focus onto the core theme of this research project, namely entrepreneurship training in rural parts of Bihar. Would greater entrepreneurship training open up new perspectives for young people in rural parts of Bihar to stay in their villages and build up sustainable livelihoods?

Entrepreneurship training is part of the agenda of India's mission for skill development. Its significance has been reinforced by the establishment of the "Ministry of Skill Development and Entrepreneurship" in 2014, which combines the responsibilities for vocational education and training that used to rest on 17 ministries, and helps to pursue the training tasks with greater efficiency. The policy for skill development and entrepreneurship defines a vision "to create an ecosystem of empowerment by skilling on a large scale at speed with high standards and to promote a culture of innovation-based entrepreneurship which can generate wealth and employment so as to ensure sustainable livelihoods for all citizens in the country" (MSDE 2015, 11). An important part to achieve this skilling mission on a *large scale*, at *speed*, and with *high standards* is played by short-term entrepreneurship training courses, which have a short duration up to six months, are free of charge, and have minimum entrance requirements that are targeted at reaching the masses of poor, especially in rural parts of the country. There are two large-scale skilling schemes relevant for the conducted field survey in Bihar, namely Pradhan Mantri Kaushal Vikas Yojana (PMKVY) and Rural Self-Employment Training Institutes (RSE-TIs), both of which will be introduced in greater detail below.

The realization of India's skilling mission requires the joint efforts of various stakeholders, which include the National Skill Development Agency, an autonomous body of the MSDE to co-ordinate the skilling initiatives, the National Skill Development Fund to raise the necessary funds for it, the National Skill Development Corporation (NSDC), a Public Private Partnership company required to catalyse the skills landscape, Sector Skills Councils, and various training providers, both for-profit and non-profit institutions, that are registered with the NSDC. The NSDC commissioned a skill-gap study encompassing 25 emerging sectors. The results are presented by division of state and sector, but the data for Bihar are not available yet. The respective skilling needs for the sectors examined within the present work will be looked at in 2.3.3.

Contribution of the private sector
At this point, special attention is given to the crucial contribution of the private sector, especially that of the Non-Governmental Organizations (NGOs)[39] in India, which makes contributions in large numbers due to the provision of aid on a large scale and a neo-liberal economic infrastructure that has resulted in a decline of government influence. This encouraged NGOs to fill the gap, which amounts to about a million NGOs in the country (Gengaiah 2016, 215). Further private players include for-profit organizations and companies that engage in skilling initiatives within their Corporate Social Responsibility. NGOs, who are usually driven by distinctive values (Nelson 1995), play a significant role in the empowerment and mobilization of the socio-economically weaker and deprived sections of the society (Chenoy 2013, Gengaiah 2016, Roy and Saini 2009). Some organizations have been promoted by the government after independence and distinctively follow a Gandhian ideology, such as the Gandhi Peace Foundation, Khadi and Village Industries Corporation, and the Association of Voluntary Agencies for Rural Development (Gengaiah 2016). But younger organizations also reflect Gandhi's values by focussing on the sustainable development of village communities, such as Drishtee, which was the main cooperation partner for this thesis.

In the following section, the two main skilling schemes in the area of entrepreneurship training, PMKVY and RSETIs, are introduced in further detail. Then follows a portrait of the training providers selected for the present work. In this approach, each organization's value-based development approach and the way they reflect Gandhi's ideology will be considered.

2.2.3.4 PMKVY-Training courses
In line with the vision of a "skilled India", the large-scale skilling scheme Pradhan Mantri Kaushal Vikas Yojana (hereafter PMKVY), which stands for "The Prime Minister's Skill Development Program", was launched in July 2015 as the flagship scheme of the new Ministry of Skill Development and Entrepreneurship (MSDE), and has been approved by the Union Cabinet for the period 2016-2020 with an allocated budget to train ten million young people (MSDE 2016). The aim of this skill certification scheme is "to enable and mobilize a large number of Indian youth to take up skill training and become employable and earn their livelihood" (MSDE 2016, 2). Thus, the focus is on both fast progress and high standards. These are ensured by the alignment of the training offer with the National Skill Qualification Framework (NSQF), as well as by a collaboration of institutions, such as the National Skill Development Corporation (NSDC) for implementing agency, Sector Skill Councils (SSCs), assessment agencies and a great number of training partners across the country. Training centres have to be registered with the NSDC and undergo an accreditation and affiliation process, during which they need to be approved by SSCs. Moreover, the training is aligned with the

---

[39] In India NGOs are "formally registered bodies" and "non-profit/public charitable organizations" (Gengaiah 2016, 214).

National Skill Qualification Framework (NSQF). The additional component of the scheme "Recognition of Prior Learning" enables the recognition of already acquired competencies in alignment with the NSQF.

The PMKVY-scheme envisages the prospect of aligning this training with the needs of the country. More than 30 sectors have been defined, of which agriculture is only one. Besides learning certain skills, the curriculum also includes further competencies, such as entrepreneurship, soft skills, and financial and digital literacy. A combination of theory and practice is fostered by operating the training with an equal number of hours in both the classroom and the laboratory. According to the official guidelines (MSDE 2016), any candidate partaking in this skill certification scheme must fulfil the minimum requirements of being of Indian nationality and being either unemployed or a school or college drop-out.[40] Each successful candidate is promised a monetary reward of 8 000 Indian Rupees, which is equivalent to approximately 100 Euros. Under the scheme, placement-objectives have also been defined, which aim for either a placement in a job or a successful placement in a self-employed activity, in which case the primary responsibility is assigned to the training centres. Thus, PMKVY makes a significant contribution in encouraging self-employment in conjunction with the development of sustainable livelihoods.

Since PMKVY was launched very recently, only a few of the selected entrepreneurs received their entrepreneurship training in the area of dairy farming under PMKVY. But multiple interviews have been conducted with managers, teachers and trainees of training centres offering PMKVY-courses, in order to obtain a better picture regarding the structural conditions of young entrepreneurs and to capture the future aspirations of the trainees. Among these are Drishtee-training centres in the Bihari districts of Madhubani, Bhagalpur and Munger.

2.2.3.5 Rural Self-Employment Training Institutes (RSETIs)

The Indian government is currently setting up "Rural Self-Employment Training Institutes" (RSETIs) in every district of the country, with the overall objectives listed as the "transformation and empowerment of rural youth for their socio-economic emancipation through training in skills development and counselling". It is hoped that these institutions will be effective in "empowering them to take up self-employment ventures, developing confidence in them and making them self-dependent" (Moodithaya 2009, 127-128). RSETIs are a follow-up model of the so-called Rural Development and Self-employment Training Institutes (RUDSETI), which have been launched rudimentarily in only a few states of the country. RSETIs are built on the same value-base as RUDSETIs, but are being established strategically to enable unemployed rural youth from below the poverty-line in every district of the country to acquire skills, and thus to play a crucial role in the mammoth task of skilling half a billion Indian youth. This is happening under the initiative of the Ministry of Rural Development in cooperation with state governments and banks. The latter of these are responsible for the management of RSETIs, which offer training in 60 agricultural as well as non-agricultural sectors, and they guarantee continued handholding with their graduates to ensure the sustainability of micro-enterprises and credit linkages after the training. The duration of these programs ranges between one and six

---

[40] This requirement prohibits any college student to partake in PMKVY-training programs (MSDE 2016, 7).

weeks. For the field survey, it was possible to visit an RSETI-institution in the Madhubani district (see below, 5.2).

2.2.3.6 The private organizations included in the field survey in Bihar, India

At this point a brief overview of the private training providers which have been visited for the field survey is given. These private organizations have in common the fact that they were established in the area of social entrepreneurship[41]. Only those where interviews with the graduates have been conducted are listed.

The social company Drishtee: developing sustainable communities

The social company Drishtee is a for-profit organization and partner of the "National Skill Development Corporation". It is focussed on developing sustainable communities. The vision-statement of Drishtee envisions "a world where all communities are empowered to achieve shared prosperity" (Drishtee n.a.). Training is only one element of this within the broadly-based development strategy. Drishtee follows a framework, portrayed by four C-words, encompassing the core areas of activity:

- **Community** – the starting point is always the village community, where connections are built and business opportunities and skilling needs are identified.
- **Capacity** – which comprises the skill training with a distinctive focus on entrepreneur-ship training according to available opportunities.
- **Credit** – which means the organization of the required capital, particularly for entrepreneurial activities.
- **Channels** – a means of building linkages between both backward and forward linkages between the village entrepreneurs and the market place.

Over the years, Drishtee has facilitated and supported a network of over 14 000 rural enterprises. It is currently working in 6 000 villages and has a strong presence in the Indian states of Assam, Bihar and Uttar Pradesh. With a particular focus on the strengthening of village communities, Drishtee follows the values of Gandhi. In the area of textile production it is engaged in the entire production chain, the making of yarn (spinning with a charkha), the handloom-based weaving, the stitching of garments, and the retail procedure. Spinners and weavers work on a self-employed basis and have been included in the survey. Drishtee has also started a village school project in the area of elementary education, following the example of Gandhi's model of basic education.

Drishtee, as part of their capacity-building initiative, is partnered with the "Bihar Rural Livelihoods Promotion Society". The "Drishtee Skill Development Center" is aiming to train 13 000 youth from below the poverty-line, 6 000 of them from Bihar, in agricultural and non-

---

[41] Just as is the case with "entrepreneurship", there are also different conceptions of "social entrepreneurship". However, there is at least a consensus that social entrepreneurship is focused on the combined achievement of both social and economic goals (Alegre et al. 2017; Singh 2016). The providers of entrepreneurship training included in the present analysis were primarily focussed on the empowerment of the deprived sections of the society, and thus reflected this characteristic.

agricultural vocations, such as wheat cultivation, paddy cultivation, stitching, masonry, sales and marketing, facility management/housekeeping, hospitality services or handicrafts, among others. Some of these training courses are set out for women.

Fateh Help Society

The Fateh Help Society is a small registered NGO which is only present in the district of Bhagalpur, Bihar. It is focussed on the empowerment of women and helps especially underprivileged children and youth. One village location with a presence of the Fateh Help Society was visited during this study, where women were trained to become self-employed in the making of hand-bags. According to the information of the unit leader, there are further units of this NGO in the area of textiles, in which entrepreneurship training is also provided, among which for the purposes of preparing yarn, stitching and weaving.

2.2.3.7 The training curricula of RSETI, Drishtee and the Fateh Help Society compared

Training curricula provide a profound basis within this dissertation's focus on entrepreneurship training as a key instrument for the empowerment of disadvantaged youth. In particular, they outline how these young entrepreneurs are being equipped. Hence, the main features of the curricula of the providers of entrepreneurship training in this dissertation will be compared.[42] In this way, emphasis is placed not so much on technical knowledge in relation to different subjects, but rather on how much entrepreneurial aspects are being considered, e.g. in the development of entrepreneurial competencies. On the basis of the differentiation between self-employment and entrepreneurship (as was discussed under 2.2.3.1), the question at hand is the following: Does the entrepreneurship training actually prepare the trainees for going on to become true entrepreneurs or not? In the modules of RSETI's entrepreneurship training programmes, the "creation of entrepreneurs and entrepreneurial attributes" are considered in the category of "most significant" (NIRD 2011, 13). In this connection, 15 entrepreneurial competencies have been identified, which can be organized within the five clusters: "achievement", "thinking and problem solving", "maturity" (which includes self-confidence), "influence", and "directing and controlling" (NIRD 2011, 13). Viewed in co-operation with Giddens' theory of structuration, competencies seem important as they are a prerequisite for competent action, and are needed for bringing knowledge into action. Thus, it can be assumed that inherent entrepreneurial competencies empower the entrepreneur not only to generate income through a self-employed activity, but also to reach the stage of entrepreneurship. At this point, it seems sufficient to mention only the first two competencies of the achievement cluster, namely "initiative" and "seeing and acting on opportunities". These two competencies are focussed upon, since they resonate with my previous considerations on the relevance of the development of personal initiative for the entrepreneur's success (see 2.1.3).

Table 2 provides an overview of the curricula of the relevant training providers arranged by subject and length. The curricula of the Fateh Help Society and Drishtee – with the exception

---

[42] Since this section deals with the analysis of different training curricula, it could have been included in Chapter 6 (Analysis). However, the analysis' main focus below concerns the way that the surveyed entrepreneurs acted in the entrepreneurship training. Thus, at this point, the training curricula will be considered as part of the enabling or constraining structural conditions in connection with the theoretical framework.

45

of dairy farming – are only based on oral information provided through the interview with entrepreneurs and training representatives. In the case of Drishtee's dairy farming curriculum, a copy in Hindi had been acquired, of which the English summary is given in Appendix I. The RSETI-curricula are found in the Appendices II-V. In the case of handlooming, no training provider was visited, and therefore no curriculum is included in Table 2. It can be emphasized at the outset that all curricula include entrepreneurial aspects. Yet, differences appear with respect to how much attention is given to these aspects. At this point, the further evaluation of these differences is limited to the curricula which have been acquired in a written form; these represent the majority of the surveyed entrepreneurs.

Table 2: Overview of the training curricula

Training Provider	Subject	Duration	Single interviews with entrepreneurs
Drishtee	Dairy farming (Appendix I)	1 month	8
RSETI	Dairy farming (Appendix II)	6 days	2
RSETI	Advanced Dairy Management (Appendix III)	21 days	No interviews conducted
RSETI	Inland fisheries (Appendix IV)	6 days	1
Drishtee	Vermicomposting	4 months	1
Drishtee	Rangoli shop management	21 days/ 1 month	5
RSETI	Beauty parlour management (Appendix V)	1 month	1
Drishtee	Charkha training	1 month	2
Drishtee	Stitching	1 month	1
Fateh	Handbag making	15 days	2

Source: Author's own draft.

In the case of the RSETI-courses, the consideration of entrepreneurial aspects is strongly identical for the three included courses. The curricula are not banded together in one distinct entrepreneurship session, yet entrepreneurial competencies are reflected throughout the entire curricula. Right at the beginning, emphasis is placed on entrepreneurial aspects by introducing the entrepreneurial competencies and the respective subject as a self-employment venture. Differences appear in the length of training. Both the RSETI dairy farming training (Appendix II) and the training in inland fisheries (Appendix VI) only last 6 days each. These curricula further include exclusive sessions on lending schemes, economics of a dairy unit, problem-solving, and a field visit from successful entrepreneurs. Furthermore, they include an interactive game (tower building) to foster the entrepreneurs' independence. If these entrepreneurial sessions are taken together, they make up at least seven sessions (out of 24) plus the tower building-exercise, which would sum up to about two full days of training, which is one third of the training course. In contrast, the training in beauty parlour management lasts for 30 days, and thus considers entrepreneurial aspects in greater detail.

Drishtee's training for dairy entrepreneurs (Appendix I), which lasts 32 days, is much more detailed than that of RSETI. Drishtee's curriculum has its own section on entrepreneurship, encompassing five days (20 hours), which includes – similarly to the basic RSETI course on dairy farming – sessions on dairy farming as a self-employment venture, different government schemes, and how to avail support, as well as cost calculation. Greater attention is given to market aspects, and how to establish customer relations, albeit not in too much detail.

However, RSETI also offers a one-month training in 'Advanced Dairy Management' (Appendix III), which covers certain topics that are not covered within RSETI's basic course on dairy farming, and which also puts greater emphasis on entrepreneurial aspects than Drishtee's entrepreneurship training. For example, RSETI's curriculum on Advanced Dairy Management puts greater emphasis on marketing management and pricing, and it explicitly considers the advantages of self-employment over waged employment, effective communication skills, and how to develop a business plan. In addition, it includes unique sessions on cattle breeding, and how to improve the milk production through it.

It can be concluded that the curricula of Drishtee and RSETI at least cover entrepreneurial aspects, and thus foster an increase not only of the specific subject-related knowledge, but also of specific knowledge regarding entrepreneurial aspects and competencies. Yet the question remains of how much of what has been taught is actually put into practice. Would these training courses also foster the development of personal initiative? As we will see below, the Analysis (Chapter 6) tries to pay attention to the distinction between the acquisition of technical knowledge and entrepreneurial knowledge/competencies. Nevertheless, NIRD (2011, 13) points to the fact that "the course module should be considered as a guideline which can be modified suitably to the local requirements". Therefore, the extent to which each curriculum is transferred into practice is dependent on the individual training centre, and ultimately on the trainer. Thus, from this brief comparison of the training curricula, no conclusions can be drawn regarding the question whether these courses actually create entrepreneurs rather than just self-employed individuals.

2.3. Enhancing livelihoods in rural parts of Bihar, India

In the last section of the theoretical framework, the previous considerations on the empowerment of disadvantaged youth through entrepreneurship training will be applied specifically to the context of the Indian state of Bihar. So far, the youth sector has been approached through the recognition of current trends within development geography from an action-oriented perspective, as well as from a pedagogical perspective, in order to consider their socialization and entrepreneurship training as an instrument of empowerment. This culminates in the guiding question of what perspectives open up for disadvantaged youth to develop sustainable livelihoods in rural parts of Bihar. First of all, it is necessary to look at the geographical characteristics of Bihar in general, and the selected target regions specifically. Then, attention will be drawn to the crucial role of community-based institutions in Bihar, in order to gain important background information on how the poor people became organized to build up sustainable livelihoods. Finally, prospects for livelihood development in the examined sectors, which are dairy farming, textiles and beauty and wellness, will be elaborated.

2.3.1 Geographical characteristics of Bihar

The goal of this section is to obtain an understanding of the agricultural conditions of the Indian state of Bihar, which is an important prerequisite for the development of sustainable livelihoods. Therefore, Bihar is looked at from the perspective of physical geography, providing an overview regarding its physical features and climatic conditions. Special attention is drawn to the target districts of Madhubani, Bhagalpur and Munger.

Agricultural-climatic zones

Bihar is part of the Indo-Gangetic Plain, which reaches from Bangladesh to Pakistan (Dayal 2013). This plain is subdivided into further divisions, one of which is the middle Gangar Plain, which captures the whole state of Bihar. It reaches from the Tarai region of Nepal in the south to the Southern plateau at Jharkhand, which is also called Chota Nagpur. This is where the Indian peninsula starts. The eastern and western boundaries of this division are not precisely defined. Based on soil, rainfall and temperature, Bihar can be divided into four agricultural-climatic zones (Prasad 2007). Zone I is the Northwest Alluvial Plain and includes the district of Madhubani. Zone II is the Northeast Alluvial Plain north of the Ganges, while Zone III is subdivided into Zone IIIA, the Southeast Alluvial Plain and Zone IIIB, the South West Alluvial Plain, which include the districts of Bhagalpur and Munger south of the Ganges. Figure 4 illustrates these agro-climatic zones.

Figure 4: District map of Bihar with agricultural-climatic zones

Source: Department of Agriculture, Government of Bihar, http://krishi.bih.nic.in/
(accessed October 26, 2017).

The river system
A highly characteristic feature of the plain is the river system, consisting of the Ganges in the middle, which divides the plateau into the North Gangetic Plain (Zones I and II) and the South Gangetic Plain (Zones IIIA+B), and many streams that enter into the Ganges from both sides, originating in the Himalayas and the Chota Nagpur Plateau respectively. This makes Bihar very prone to floods, as the whole plain is divided into so-called 'doabs', which is the land between two rivers (Dayal 2013). The Zones I and II are in general more prone to flooding, and Zones IIIA and B are more prone to draught. The main rivers which originate in the Himalayas and which pass by or through the target-district of Madhubani on course to the Ganges are the Baghmati, which rises north of Kathmandu, Nepal, which passes the Madhubani district in the west, and which converges with the Kosi river before entering the Ganges; the Kamla, which crosses Madhubani and enters into the Baghmati south of Madhubani; and the Kosi, which passes along the eastern border of the district. The Ganges runs south of the district of Munger and crosses the target-district of Bhagalpur in the south of the district. Therefore, some of the villages visited in Bhagalpur are south of the Ganges, and some are north of it. There are some smaller streams draining into the Ganges from the south and affecting the district of Bhagalpur. In the dry season, these streams are sandy water courses, but in the rainy season, particularly after heavy rain, they can turn into hazardous streams of considerable size, bringing rapid floods (Dayal 2013). The biggest among them is Chandan. South of the city of Bhagalpur it has a delta-shaped appearance and enters the Ganges by several mouths. The visited village of Puraini, which is part of the handloom cluster of Bhagalpur, had been flooded by the Chandan system in the past. The Kamla river is also subject to frequent floods, but this is apparently not a big threat to the visited entrepreneurs in the Madhubani district.

The extensive river system determines the quality of the soil, which is characterized as alluvial (Dayal 2013). Its fertility is determined by proximity to a river. Older alluvium is called Bangar, which is more upland and usually not as fertile, but less prone to floods. Newer alluvium is called Khadir and is very fertile. Khadir plains are nearer to the river and flood-prone, sometimes in a former riverbed, which run dry due to the change of the river course. Due to the different levels of productivity and relevance for tax collection, villages have been historically classified in Khadir, Khadir-Bangar and Bangar.

Climatic conditions
The overall climate in Bihar is hot and sub-humid (NICRA, n.a.). In the hot season from March to May the average temperature is 35-40 degrees Celsius. The rainy season is from June to October and the cold season from November to February (Dayal 2013). Table 3 provides a brief profile of the annual rainfall within the target-districts of Madhubani, Bhagalpur and Munger. Whereas the average rainfall of the main monsoon shows only slight differences between these districts, the summer season reveals huge differences, with Munger being the driest region. Table 4 shows the degree to which the respective districts are prone to drought, flood, and hot or cold waves. This can have a severe impact on the development of sustainable rural livelihoods.

Table 3: Annual rainfall in the target-districts (in mm)

District	Southwest Monsoon (Jun-Sept)	Northeast Monsoon (Oct-Dec)	Winter (Jan-Feb)	Summer (Mar-May)	Annual
Madhubani	984.8	72	24.6	103.6	1185
Bhagalpur	992	97	26	93	1208
Munger	952	115	31	45	1143.1

Source: NICRA (n.a.), data compiled from the respective district-wise agriculture contingency plans of NICRA.[43]

Table 4: Vulnerability of the target-districts to climatic events

District	Drought	Flood	Hot wave	Cold wave
Madhubani	regularly	occasionally	occasionally	occasionally
Bhagalpur	occasionally	regularly	regularly	regularly
Munger	occasionally	occasionally	occasionally	occasionally

Source: NICRA (n.a.), data compiled from the respective district-wise agriculture contingency plans of NICRA.

Agricultural conditions

Bihar is generally very flat. It has a very poor forest cover of only 6%, but it is richly cultivated (Dayal 2013). In Bihar, traditional types of cropping with mixed cropping are common (Sinha 2009). For the sake of irrigation, canals are predominant in the districts of Munger and Bhagalpur. In the Madhubani district are found mostly irrigation tanks as part of the ancient tradition (NICRA, n.a.). In the agriculturally more developed Zones I and IIIB, a higher area of land is under irrigation, which is why the grazing land declined. As a consequence, the density of cattle is comparably lower, but the density of buffalo higher than in other states of Bihar (Singh et al. 2010).

There are three agricultural seasons in all of Bihar (Sinha 2009):

- In the Bharai season, quickly maturing crops such as rice, millets, maize and jute (in the north of Bihar) are sown in early June, prior to the monsoon showers, and harvested between August and September.
- In the Aghani season paddy is typically sown in all of Bihar when the monsoons start, and harvested in November/December.
- In the Rabi season, wheat, barley, pulses and oilseeds are sown before the Aghani crops are harvested, and are harvested in March. Rabi-crops largely depend on the availability of irrigation facilities, and therefore are more harvested in canal areas.

Table 5 provides an overview of the most common crops of the target districts. Except for dairy farming, none of the agricultural activities had been included in the survey, yet many of the

---

[43] National Innovations on Climate Resilient Agriculture (NICRA), a research based network project established to enhance agricultural resilience, and launched by the Indian Council of Agricultural Research (ICAR) in February 2011.

50

surveyed entrepreneurs were farmers, as agricultural diversification is a common livelihood strategy. Hence, this information provides important background knowledge regarding the understanding of how to improve rural livelihoods, which will be considered in the following two sections (2.3.2 and 2.3.3).

Table 5: Major crops of the target-districts

District	Major crops	Further crops	Horticultural fruits
Madhubani	rice, wheat, maize	pigeon pea, rabi pulses, oil seed/mustard	mango, guava, banana
Bhagalpur	maize, rice, wheat	mustard, barley	mango, guava, litchi, lemon, banana
Munger	rice, wheat	maize, mustard	mango, guava, litchi, lemon, banana

Source: NICRA (n.a.), data compiled from the respective district-wise agriculture contingency plans of NICRA.

2.3.2 The importance of community-based institutions

Social capital is a driving factor for entrepreneurs in any field (Burt 2000; Prajapati and Biswas 2011). For rural entrepreneurs in India, and especially those from lower social sections of the society, not only training but also the availability of a social network is crucial, for which community-based institutions play a significant role. In light of a wide range of institutions and schemes in India offered by the government and banks for female entrepreneurship, microcredits and the like, one may easily lose one's orientation. Therefore, this sub-section starts with an overview of three significant institutions, which have developed in India throughout the whole country, and which are crucial for individual entrepreneurs to grow more connected, namely cooperative societies, farmers' clubs and self-help groups (SHGs). A wide range of Indian scholars have focused on the role of Indian cooperatives – and SHGs in particular – in connection with empowerment and entrepreneurship. Here, let it be mentioned by way of an example of women's empowerment through entrepreneurship the collected volume of Rathakrishnan (2008), which includes an exclusive section on SHGs as a means of women's sustainable entrepreneurship with seven individual contributions, or the collected volume of Vasanthagopal and Santha (2008), which has a detailed overview of schemes both by the government (e.g. Scheme of Assistance to Women Co-operatives), and private institutions, e.g. financial institutions. After this, the National Rural Livelihoods Mission is introduced as an important medium for the empowerment of the rural poor, which in Bihar was implemented as the Bihar Rural Livelihood Project. These supportive theoretical mechanisms are important for the detailed overview regarding the prospects for the development of rural livelihoods in the examined sectors, which follows in 2.3.3.

Evolution and impact of cooperative societies, farmers' clubs and SHGs
In India cooperative movements have a long tradition, whose number has consistently increased since independence. While in 1984/85 over 300 000 cooperative societies existed in India, its number has doubled by 2009/10 with more than 600 000 cooperatives and 260 million mem-

bers, covering 98% of Indian villages (National Cooperative Union of India 2012, 33). Cooperatives have a strong influence on the dairy sector, with more than 140 000 dairy cooperatives in 2009/10, who process the biggest portion of milk. The 'white revolution' under the motto *Operation flood*, a rural development program for the advancement of milk production which took place in three phases between 1970 and 1976, was based on village milk producers' cooperatives and brought structural changes in favor of small dairy farmers, *inter alia* through the creation of a national milk grid (Scholten 2010). The key player that fostered the white revolution in India was Amul, which is the biggest cooperative society in India.

Farmers' clubs are associated with the National Bank for Agriculture and Rural Development (NABARD) and function as informal forums for local farmers who share mutual benefits from banks, which are assisted by NABARD. Upon the inception of NABARD in 1982, the Indian Government carved out a way to facilitate credit flow to the rural population, fostering the development of the agricultural sector, as well as the cottage and village industries. Since 2005 farmers' clubs are covered under the Farmers Club Programme of NABARD.

SHGs are informal savings collectives with mutual accountability, often being granted with microcredits from a bank (Islam and Imam 2011). In India, a SHG bank-linkage program has been in place since 1992, which, since its conceptualization, has experienced dramatic physical and financial growth. In 2009 more than six million SHGs had saving bank accounts, with a total savings of more than 900 million Euros, and more than four million had credit-linkages with banks, from which an estimated number of 70 million poor Indian households benefited. About 80% of the SHGs were women groups (Mohanty 2013, 63). Yet as is stated by the NSSO survey in 2003, only 50% of all farmers had access to any source of credit (Mohanty 2013, 56). SHGs also play a significant role in India's National Rural Livelihood Mission (NRLM), and therefore this aspect will be further introduced in connection with the next section, the development history of NRLM.

Despite these positive trends regarding the impact of community-based organizations, Singh et al. (2010, 1) notes – based on the information of the NSSO (2004) – that, in the state of Bihar, only about 1% of the farmers are associated with farmers' organizations (including SHGs), and about 5% of farm-households are members of co-operative societies.

Promotion of rural livelihoods in India

In 1990 the Indian government launched the program "Swarna Javanti Gram Swarozgar Yojna" (SGSY) as a group-focussed self-employment program for the poor. In this a shift took place from an individual-oriented approach to a group-based approach, with SHGs of rural families below the poverty line acting as units of assistance. This program was aimed at linking the key elements of capacity building, infrastructure and technology provision, access to credits, and marketing. Based on recommendations of the Prof. Radhakrishna Committee (MoF 2007) SGSY has been restructured into the "National Rural Livelihoods Mission" (NRLM), which is also known as "Ajeevika". This program was formally launched in 2011 to improve the shortcomings of the former program, such as the unequal mobilization of the rural poor, weak capacity building, low credit mobilization and a lack of dedicated professionals to implement the program (MoRD 2014). In 2015 the program was revamped and named Deen Dayal Antoyodaya Yojana – National Rural Livelihood Mission (MoRD 2017). MoRD defines SHGs as "homogenous groups of 5-20 women that function on the principles of mutual cooperation and collective action. These SHGs are federated at the village/Gram panchayat, cluster and block level" (MoRD 2017, 17).

In Bihar NRLM was jointly implemented in 2007 under the name "Bihar Rural Livelihood Project" (BRLP) in cooperation with the World Bank, the Government of Bihar and the community-based organization "Bihar Rural Livelihoods Promotion Society", which is also known as "Jeevika" and serves to represent the poor (Behera et al. 2013). This program is targeted at small and marginal farmers and has designed a "single-window system at the doorstep" to help them to gain access to their basic needs and meet their requirements. This system integrates initiatives such as bulk purchases of seeds, customized extension services, and access to low-cost-credits. Farmers are organized into SHGs of 10-15 individuals. These in turn are organized in "Village Organizations", which comprise of about 20 SHGs, which again form "Cluster Level Federations", consisting of about 20 Village Organizations. Through this structure a pro-poor investment climate is being created, enabling the poor to obtain low-cost-credits through commercial banks. Additional elements are the implementation of the "System of Crop Intensification" to improve cultivation methods and productivity and "Zero Budget Natural Farming", a method of organic farming which allows a farmer to sell his products at higher rates. A total number of cumulated credit-flow from banks of about 25 million Euros and a cumulative household financial turnover of about 190 million Euros between 2007 and 2013 stands as proof of the significance and worthwhile impact of this program on the livelihood development of the rural poor in Bihar.

2.3.3 Prospects for livelihood development in the examined sectors

Now focus will be turned to the occupational groups included in the present study. As young entrepreneurs are considered as occupying the area of tension between agency and structure, this sub-section outlines the structural challenges which young entrepreneurs from socially deprived backgrounds are confronted with in the surveyed sectors, namely dairy farming, textile production and beauty and wellness, more specifically in opening a small ladies' shop or a beauty parlour.[44] In this way, attention is drawn to the prospects for livelihood development in these respective sectors, for which the previous insights on the significance of community-based organization are relevant. Hence, the importance of SHGs for the empowerment of the poor will be shown in the different sectors. In addition, some challenges will be addressed, which appear in connection with the growth of micro-enterprises within markets at the 'base-of the (economic) pyramid', which is the case in the selected target regions in rural parts of Bihar.[45] Economists have characterized these markets as informal (Webb et al. 2013) and as "unserved or underserved by the large organized private sector, including multinational firms" (Prahalad 2014, 6). Nevertheless, Prahalad points to the potential for the development of these

---

[44] In order to reduce the complexity of this research, at this point two occupational groups, namely inland fishery and vermicomposting, will not be addressed in greater detail, as only one interview for each was conducted. The vermicomposting entrepreneur came from a wealthier background and had this activity only as a side-business, while the interviewee in inland fishery did not pursue her activity any longer.

[45] See Prahalad 2014 and his quotation in the opening of this chapter.

markets, which would "create millions of new entrepreneurs at the grass-roots level – from women working as distributors and entrepreneurs to village-level micro-enterprises" (Prahalad 2014, 6). With this in mind, the question arises how markets within the surveyed sectors could be developed through micro-enterprises at the base of the pyramid, which is part of a controversial debate: do these micro-enterprises, which are often home-based and run by a single individual, merely serve for the subsistence of a family, or do they actually fulfil the greater purpose of job creation and economic growth? (Mandelman and Montes-Rojas 2009; Webb et al. 2013) This prevalent question ought to be kept in mind, as the surveyed sectors are being looked at. Information has been gained from the skill gap analysis conducted by KPMG[46] (n.a.) for the planning periods 2013-17 and 2017-22, and from further reports and studies that have been undertaken.

2.3.3.1 Dairy farming

India is the country with the world's largest milk production, so dairy farming is one of the most important agricultural sectors and an important source of rural employment, especially for small and marginal farmers (Shah 2006). Within the present survey, dairy entrepreneurs make up the biggest group, varying from the very poor to the very rich. For the people from lower social classes this activity is closely connected to the caste system, such as in the village of Afzalnagar in the Munger district, Bihar, which has been visited for the survey. There, people belonged to the Ahir caste-group, a traditional peasant-pastoral community, which means that they have been engaged in it for many generations.[47] In connection with perspectives of empowerment through entrepreneurship training, structural limitations have to be discovered at various levels, from the national and regional level which has to do with market trends and skill requirements down to the individual level which has to do with the socio-cultural challenges of small-scale dairy entrepreneurs.

The poor people engaged in dairy farming face immense structural challenges. The Agriculture Skill Council of India (ASCI, n.a.) reveals that, in India, while more than two thirds of the rural population engage in animal husbandry, small and marginal farmers own over 60% of all milk animals. Thus, the average number of cows is very low, which makes it hard to make a living from it. 69% of people employed in dairying industry belong to socially disadvantaged groups such as scheduled castes (SC), scheduled tribes (ST) and other backward castes (OBC). However, it is very advantageous that this line of work is not dependent on seasonality, providing a stable source of income throughout the whole year, while crop production provides employment to landless, small and marginal farmers only for 90 to 120 days a year (ASCI, n.a.).

In spite of the fact that the milk consumption is expected to grow by a rate of 4% by 2022, the skill gap analysis conducted by the KPMG projects the employment number for animal production in agriculture for all of India to remain unchanged at 13.9 million for the five-year planning periods of 2012, 2017, and 2022 (KPMG, vol. 1., n.a.). The dairy sector holds opportunities for highly skilled personnel with a diploma or higher educational degrees, as well

---

[46] KPMG Advisory Services Pvt Ltd, a globally operating network of professional firms, which offers among other things services to governments, has been engaged with the creation of the skill gap analysis by the NSDC, based on independent research conducted by KPMG.

[47] Further description of the Ahir is available in the Encyclopedia Britannica (2018).

as for people skilled through short-term vocational training courses for ground support with a direct focus on the dairy farmer. In considering prospects for empowerment, attention is now given to opportunities through cooperatives and SHGs as well as to training opportunities. The Indian milk production industry is very fragmented. 22 million tons of milk are processed annually by the unorganized sector, and only 13 million tons by the organized sector. The biggest portion of milk is processed by dairy cooperatives (KPMG, vol. 1, n.a.). In the dairy sector SHGs are the main avenue for availing microcredits. Feroze and Chauhan (2010) undertook a study into the performance of dairy SHGs in the Indian state of Haryana. Their study is based on a sample of 60 members from 30 SHGs in Haryana. A performance index has been created based on various indicators, e.g. savings, credits, repayment and income generating activities, which revealed that the performance was poor for 22%, average for 47%, and good for only 30% of those surveyed. SHGs successfully availed microcredits of about 200 000 Indian Rupees[48] on average in the study period 2006/2007, which follows the credit to saving ratio of 4:1, as was recommended by NABARD. This example shows the direct advantages of SHGs for the individual. Nevertheless, this instrument is most advantageous if education and training is also pursued. Roy (2013) explores connections between livestock and the development of dairy farming, and points to the fact that education, taken together with SHGs, continues an "facilitating factor in realising higher performance via better understanding of the mechanism involved in the formation and functioning of SHGs and practicing scientific dairy farming" (Roy 2013, 313). Feroze and Chauhan (2010) also point to the potential of skills training in the area of dairy farming and its positive effects on the performance of dairy entrepreneurs. The KPMG report points to the significant demand of skilled manpower, which is expected for both experts with higher educational degrees, from Bachelor holders up to PhDs, and graduates of short-term vocational training courses who are focussed on ground support of farming activities. ASCI (n.a., 8) reveals that about 92% of workers in the dairy industry engage in the primary sector, i.e. farming and allied activities, which in addition to cattle- and goat-rearing for milk production includes breeding, ranching and grazing, but only 1.2% focus on the processing of milk products, such as the manufacturing of butter, cheese, milk powder, ice-cream, and the like. This raises the question for a large number of young dairy entrepreneurs, who can hardly make a living from dairy farming alone, whether a skill upgrade in such processing activities would be a worthwhile option to expand their scope and increase their income base.

2.3.3.2 Textiles

The textile sector contributes significantly to India's economic growth and provides employment and sources of livelihoods for specifically poorer sections of the society (Sarmistha 2015). This sub-section outlines challenges and opportunities for entrepreneurs engaged in the production of garments. There are three successive stages of production in the making of garments, which include spinning, weaving and sewing. Although the present survey included the whole spectrum of the production chain, at this point only the handloom sector in Bhagalpur, Bihar, will be looked at in further detail. Information has been drawn from two different studies, which give profound insights into the scope of action and structural challenges for self-employed

---

[48] Approximately 2 500 Euro.

handloom workers and provide recommendations for their empowerment. Firstly, a diagnostic survey was carried out of the Infrastructure Leasing and Financial Services Limited (ILFS 2007), which had been mandated by the Department of Industries, Government of Bihar. This survey was recently conducted[49] among 440 weavers from various villages in Bhagalpur and other districts of Bihar, with the goal of preparing of a business plan. Secondly, an article of Sarmistha (2015) was published on the rural handloom textile industry in Bihar, which was part of a larger research project on the impact of globalization on the informal sector. She draws on her own findings based on a study in the weaver's cluster in Bhagalpur in 2005. Due to the strong degree of mutual dependency, conclusions can be drawn regarding the situation of spinners as well.

Although the handloom sector is in decline, it still provides employment to around 4.3 million workers in India (around 37 000 of which come from Bihar), who mostly come from lower social classes, such as scheduled castes, tribes, and other backward groups. They usually engage in handlooming on a small scale, which is barely sufficient for a living (Third Handloom Census of India 2009-10, MoT 2010). Handloom workers are organized into clusters. The Indian government supports the development of such clusters through different schemes, such as the Comprehensive Handloom Cluster Development Scheme, under which Mega Clusters are being developed with the aim "to empower handloom weavers and build their capacity to enhance competitiveness of their products in the domestic as well as global market in a sustainable and reliant manner" (MoT 2017a, 3). In 2015 India's Prime Minister Narendra Modi established the *India Handloom Brand* to ensure quality and better promotion and marketing of the handloom-based products.

In Bihar, the Indian Government has taken up Bhagalpur as Mega Cluster in the Union Budget 2014-15, covering the districts of Bhagalpur and Baka (MoT 2017b). Here, the vast majority of handloom workers lives in villages nearby the city of Bhagalpur and keeps up this occupation as a home-based business, within a long family tradition that is passed from one generation to the next. The city of Bhagalpur at the southern riverside of the Ganges, which is famous as the silk city, is associated with the traditional handloom industry, which is over 100 years old, and which is especially known for the Tassar silk (Sarmistha 2015).

There are various challenges that threaten the handloom industry, which need to be considered in relation to the question of empowerment of the handloom workers. Industrialization, which took off in the 1980s, almost led to the destruction of the handloom sector, as traditional spinning and weaving activities had been replaced by power-looms. According to Sarmistha (2015, 113) on a handloom only eight to ten meters can be produced a day, whereas on a power-loom 20 meters can be produced a day. There are typically three workers needed for two handlooms. According to ILFS (2007, 16), in Bihar the income-level of this male-dominated occupation is very low. 56% of handloom weavers (from a total sample of 440) have a monthly income between 1 000 and 3 000 Indian Rupees[50], while only 34% are above this margin, and 10% are below it. In addition, handloom workers lack market linkages. ILFS (2007,

---

[49] The report itself provides no information regarding the date of the survey, but the link to the pdf document under the homepage of the Department of Industries, Government of Bihar, indicates April 24, 2007 as the date of publication.

[50] Which is roughly between 15 and 40 Euros.

29) found that in Bhagalpur a majority of 68% of weavers (from a sample of 60) sell their products in local markets, about 18% do so to government institutions, and only 14% do so in regional markets. Their weak economic status coupled with a lack of education and bar-gaining power leaves them dependent on middlemen and vulnerable to exploitation. According to Sarmistha (2015), the majority of the poor handloom workers depends on so-called *Mahajans*, who are moneylenders or traders, who control the handloom market and profit from it disproportionately. ILFS (2007, 29) reveals a failure of financial support through banks, as in the Bhagalpur district where only 19% of handloom workers receive credits from banks, 34% obtain it from money lenders, 25% obtain it from family members or friends, and the rest remains without any credit. Only 50% of them owned the handlooms, while the rest were working on looms provided by traders (ILFS 2007, 15). Sarmistha (2015) also points to the poor health condition of handloom workers. Due to long working hours with bare hands, a high proportion of them has been infected by tuberculosis or afflicted by infections of the hands or lungs, which causes a high migration of skilled adult weavers out of the villages, many of whom end up as unskilled workers in other sectors. At the same time, ILFS (2007, 23) found that younger people perceive weaving as a potential career, but one that is not as profitable. Only one third of the surveyed weavers (from a sample of 60) had gone through any type of skill-upgradation programme. The Weavers Training Centre at the Puraini village in the Bhagalpur district was found to be in a bad condition, with old equipment and poorly qualified trainers (ILFS 2007).

All these structural limitations raise the question of what it means for a handloom worker to be empowered and what role local training providers, NGOs and community-based federations can play to foster the process of empowerment. Sarmistha (2015) lists as necessary interventions the release of handloom workers from their dependency on exploitive middlemen, for which an effective infrastructure for distribution and marketing and supply chain need to be in place, the upscaling of technology and a thorough improvement of the health system. Sarmistha and ILFS both refer to the positive effects of the weaver's federations and SHGs, in which its members mutually gain and benefit through access to credits from banks.

2.3.3.3 Beauty and Wellness

Finally, attention is given to the services sector, most specifically to the beauty and wellness sector, in which India ranks among the top five globally. With a compound annual growth rate of 18.6% this sector has been identified as one of the 25 growth sectors in India (KPMG, vol. 4, n.a.), and it thus holds great potential for the empowerment of socially deprived groups, particularly for women. The continuous growth is fostered by a burgeoning middle class with an increased awareness and demand for beauty- and wellness-related products. This growth trend goes hand-in-hand with an increased focus on quality of services, which raises the demand for a skilled workforce in this segment. The required workforce is estimated in 2017 to have been 739 000 and for 2022 to be 1 427 000 (KPMG, vol. 4, n.a., 21). The sector is highly unorganized. More than 70% are informal workers, and many of this group are rural entrepreneurs from lower income groups.

The sector is further subdivided into distinct sub-sectors based on unique skill requirements (KPMG, vol. 4, n.a.). Relevant for the present survey are the two sub-sectors *beauty centres*, which correspond to the visited beauty parlours, offering cosmetic treatments exclusively to woman, including pedicure and manicure, and *product and counter sales*, the sale of beauty and cosmetic products at salons and retail outlets. The latter may comply with the visited Rangoli outlets of Drishtee, which are all-women multi-purpose shops. Hardly any research could be found on women empowerment in these segments within the beauty and wellness sector. Apart from the KPMG report, relevant insights on the subject are drawn from two sources: on the one hand from a very recent report of the Hindustan Times on Drishtee's Rangoli outlets, which complement the collected data (Sopam 2017), and on the other hand from Tarakumari (2008), who undertook a survey into socio-economic issues related to women working as entrepreneurs by running a beauty-parlour. This survey covered 45 beauty-parlours in an urban setting, the city of Visakhapatnam in the Indian state of Andhra Pradesh. The structural conditions in these urban areas are, without question, very different from those in rural areas, yet the study provides a good overview on general market trends and some valuable insights on marketing strategies and determinant factors of success. This study thus permits us to attempt to make them applicable to the rural context of Bihar.

Tarakumari (2008) considers the location, the accommodation and the equipment as factors of success. In rural areas, the location may be one of the biggest challenges, as the demand for beauty related services may be very limited, leaving only few opportunities for growth. But due to a low capital base, the entry barriers for opening a beauty parlour are relatively low in urban areas. This may be even more so in rural areas, as the issues of accommodation and equipment are smaller obstacles, since one is often able to open a beauty parlour at home, an undertaking which only requires the availability of at least a separate room, and the minimal costs of equipment needed to get started.

In the case of a Rangoli shop the starters capital is 25 000 Rupees[51] and the monthly income was reported to range between 3 000 and 5 000 Rupees[52] (Sopam 2017), which is within the normal range for rural parts of Bihar.[53]

Taken in connection with the present paper's focus on youth and the skilling initiatives and cooperative movements, the question arises what opportunities for empowering socially disadvantaged youth become open through short-term entrepreneurship training courses in the area of beauty and wellness, and what significance the SHGs have in this area. For the visited training locations in Bihar the duration of the training in preparation for the establishment of a beauty parlour ranged between one and two months, while the Drishtee entrepreneurial training for woman to open a Rangoli shop was a month.

In order to encourage rural entrepreneurs, the skill gap report (KPMG, vol. 4, n.a.) recommends government support with respect to retail space and infrastructure for the establishment of beauty and wellness retail outlets in rural areas. With respect to the SHGs, the recommendation is to promote vocational training in the beauty field for SHGs. The KPMG report mentions as weaknesses of the beauty and wellness sector a lack of standardization of training programs, leading to a mismatch between the acquired skills and market demands, and a lack of penetration of the vocational education and training programs that are offered, which may

---

[51] Between 300 and 350 Euros.
[52] Between 40 and 70 Euros.
[53] The income per capita for Bihar was only about 35% of the national average in 2015-16 (GoB, n.a.).

result from an inadequate awareness of possible career prospects, but also from a low overall attractiveness of the sector. As their report writes, "a large portion of the workforce choose this path due to lack of options" (KPMG, vol. 4, n.a., 45). The KPMG report also refers to the difficulties faced by candidates with a low capital base to secure formal loans for entrepreneurial activity and recommends to "promote vocational training for SHGs in the field of beauty and wellness", and to "provide common infrastructure facilities on a plug and play model for rural entrepreneurs to offer their services" (KPMG, vol. 4, n.a., 49). The latter ploy is certainly fulfilled in the case of Drishtee's Rangoli initiative, which is supported by a Japanese sponsor (Sopam 2017), and which is an example of a sound infrastructure, where Drishtee provides the training, the start-up support including the start capital in form of a loan, and continuous assistance of the women entrepreneurs. The present survey ties in with this question of how these women could become more empowered to improve their structural conditions.

3. The Research Question

It is now time to focus on the general research question (3.1), which constitutes the key element of the present dissertation, towards which the theoretical framework has been oriented. This general research question will be further subdivided into specified research questions (3.2), according to which the further breakdown of the theoretical framework has in turn been undertaken. This breakdown is illustrated in Figure 5, which summarizes the key findings from the previous chapter's theoretical considerations. It combines the two main parts of the theoretical framework, firstly a consideration of how the youth sector gains access to sustainable livelihoods from an action-oriented perspective, and secondly a consideration of the socialization of youth from a pedagogical perspective. Young entrepreneurs find themselves in a process of transition into adulthood, in which they are placed within an area of tension between agency and structure (Giddens 1984). In this situation, they are confronted with the challenge of developing a livelihood strategy in compliance with their personal motivations and perceptions.

Figure 5: Developing a livelihood strategy within the area of tension between agency and structure

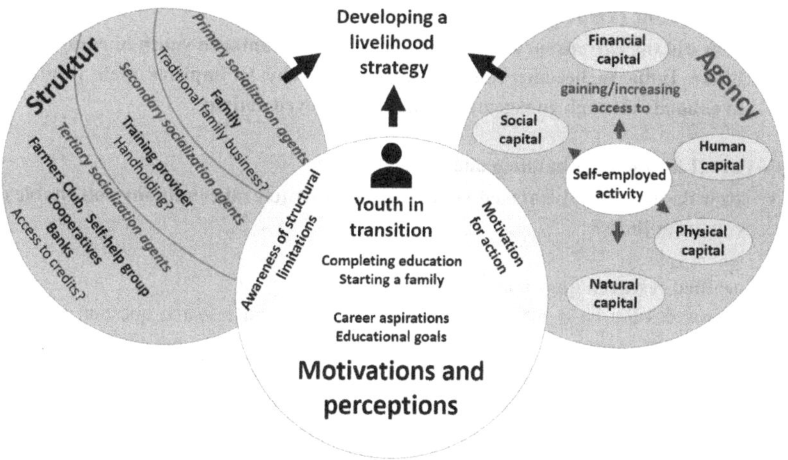

Source: Author's own image.

3.1 The general research question (GQ)

The general research question deals with the core theme of this dissertation, namely the duality of agency and structure. On the one hand is the agency of disadvantaged youth from rural parts of Bihar, India, the subject of this dissertation's research, who are viewed as self-determined actors. Their agency can be expressed through their entrepreneurial activity, and their goal – in

© The Editor(s) (if applicable) and The Author(s), under exclusive
license to Springer Fachmedien Wiesbaden GmbH, part of Springer Nature 2020
T. Aberle, *Entrepreneurship Training in Rural Parts of Bihar/ India*, Perspektiven der Humangeographie, https://doi.org/10.1007/978-3-658-30008-1_3

compliance with the "sustainable livelihoods framework" (DFID 1999) – is to gain or increase access to productive livelihoods. On the other hand is the structure, which is reflected by structural conditions, that are both constraining or enabling for these entrepreneurs. The entrepreneurship training is the object of this dissertation's research as a key instrument for the empowerment of disadvantaged youth. Therefore, the question at hand is how entrepreneurship enables the young people involved in this study to cope with structural conditions. As the research design is built up in two parts, namely the evaluative qualitative text analysis and the type-building analysis, the general research question is also divided into two parts. The first question is a guiding question for the evaluative part of the analysis, and considers the entrepreneurship training in connection with the empowerment of disadvantaged youth. More specifically, it refers to the training effects with respect to their ability to improve productive livelihoods through a self-employed activity. In the course of the evaluative text analysis, *empowerment* will be itemized in its different quantifiable dimensions, such as personal, economic and sociocultural empowerment. But over and above these different dimensions, the ultimate goal is to identify different coping strategies of the young entrepreneurs. Therefore, the second part of the general research question, which is oriented at the type-building analysis, will inquire into the enabling or constraining structural conditions, and will build on the insights of the first part of the question.

GQ – Part I for the evaluative qualitative text analysis:
- **How did the entrepreneurship training help disadvantaged youth in rural parts of Bihar, India to become empowered, and thereby to improve their productive livelihoods through engagement in a self-employed activity?**

GQ – Part II for the type-building analysis:
- **How do these disadvantaged youth cope with structural conditions that enable or constrain them?**

3.2 Specified research questions (SQs)
Figure 5 provides the basis for the subdivision of the general research question into eight specified research questions. The first six SQs are guiding questions for the evaluative qualitative analysis and are grouped in accordance with the three cycles, namely the cycle of agency, the cycle of structure and the cycle of personal motivations and perceptions. The final two questions look beyond these three areas and point to the final result of the study. SQ7 is a guiding question for the type-building analysis, and SQ8 regards recommendations for the training providers and the involved stakeholders, which can be derived from the study.

Specified research questions regarding the agency of young entrepreneurs
The first two specified research questions relate to the agency of disadvantaged youth, which – in compliance with the "sustainable livelihoods framework" (DFID 1999) – is to gain or increase access to the five capital assets: human, financial, physical, natural and social capital. These are guiding questions for the evaluative text analysis. The first question asks about the application of specific training inputs to their business activity, which includes both technical

knowledge related to the specific subject and managerial knowledge, which fosters the development of entrepreneurial competencies. The second question considers the role of the development of personal initiative in gaining access to these assets. By the development of personal initiative is included any proactive behaviour that goes beyond the sole application of training inputs or simple adherence with the given structure. Since SQ2 not only relates to one dimension of empowerment, but to all three dimensions – that is, personal, economic and socio-cultural empowerment – in the analysis (Chapter 6), it will be considered after the category-based analysis.

- **SQ1: How did the graduates manage to utilize specific training inputs of the entrepreneurship training course, in order to gain or increase access to various capital assets?**
- **SQ2: What role does the development of personal initiative play in gaining or increasing access to various capital assets?**

The supporting hypotheses for SQ1 and SQ2
Since these research questions are incorporated into the theoretical framework, it seems appropriate to specify with hypotheses how they are intended to be understood. SQ1 deals with the process of learning in connection with Giddens' theory of structuration, and more specifically his understanding of the different levels of consciousness. Under "the duality of structure" (2.1.2), the learning process was outlined, which takes place during the entrepreneurship training, and through which the degree of discursive consciousness of the competent actor increases. It was said there that, as a concomitant effect of the learning, the amount of reflexive knowledge, which allows for the reflexive monitoring of action, increases as well. Hence the following hypothesis is formulated to supplement SQ1:

Hypothesis for SQ1:
- **It is possible to apply reflexive knowledge, which has increased through the entrepreneurship training, to the self-employed activity in such a way that the agency would *thicken* over time, and thus the capability to cope with structural limitations would increase as well.**

However, the transfer of reflexive knowledge to practice does not happen automatically. In alignment with Frese's elaborations on personal initiative and the reactive strategy (2.1.3), where he concluded that "there is only one pathway to success and that is through actions" (Frese 2000, 162), the following hypothesis, which directly builds on the previous one, is formulated:

Hypothesis for SQ2:
- **The productive application of reflexive knowledge to the self-employed activity, and the ways that enabling or limiting structures are coped with, are significantly dependent on both reactive and proactive behaviour patterns.**

The next three research questions relate to the structural conditions of young entrepreneurs and is also part of the evaluative text analysis. In dealing with the process of socialization that youth undergo as they transition into adulthood, these questions address the impact of different socialization agents on the business activity. Now, it would lie beyond the scope of the study to consider the whole range of socialization agents mentioned in Figure 3 (under 2.2.2.3). During the first process of coding and defining the subcategories, it turned out that three of them play a key role regarding the success of the entrepreneurs, and therefore they have to be considered as part of the structural set-up. These three subcategories consist in the family as primary socialization agent, the training provider as secondary socialization agent, and community-based organizations, such as cooperatives, farmers' clubs and self-help groups as tertiary socialization agents.

- **SQ3: What influence did the family as primary socialization agent have on young entrepreneurs to become involved in the self-employed activity?**
- **SQ4: How do the graduates of the entrepreneurship training benefit from the assistance and the connections of the training provider?**
- **SQ5: How do young entrepreneurs benefit from social networks with cooperatives, farmers' clubs, self-help groups, and the like?**

Specified research question regarding the motivations and perceptions of young entrepreneurs
SQ6 and its associated hypothesis are concerned with the compliance of the entrepreneurial activity with the personal educational and professional goals. This takes into account the dynamic stage that these young entrepreneurs are in as they transition into adulthood, in which they are confronted with other pulling factors that compete with their self-employed activities, such as pursuing a high-prestige job in other fields.

- **SQ6: How does the entrepreneurial activity of young entrepreneurs comply with their educational and professional goals?**
- **Hypothesis for SQ6: For most of the young entrepreneurs, the self-employed activity is, due to their higher educational achievements and personal career aspirations, only a Plan B.**

Specified research question regarding the development of livelihood strategies
SQ7 is a guiding question for the type-building text analysis and considers the different livelihood strategies that young entrepreneurs develop in order to cope with structural conditions. These are oriented at the three livelihood strategies identified by Scoones (*agricultural intensification/extensification, livelihood diversification* and *migration*), which depend on the individual's particular livelihood portfolio (Scoones 1998, 10; see under 2.1.4.2).

- **SQ7: What livelihood strategies do young entrepreneurs develop in relation to their occupational group, age, gender, social background and educational status, in order to cope with structural conditions that enable or constrain them?**

In connection with Scoones' (1998) identification of livelihood strategies (2.1.4.2) and Frese's (2000) elaboration of the reactive strategy (2.1.3), the following hypothesis is formulated, which ties in with the previous ones:

Hypothesis for SQ7:
The ways that entrepreneurs manage to "thicken" their agency or not depend on the following livelihood strategies, which they often chose in combination as part of their livelihood portfolio:

- Some entrepreneurs proactively thicken their agency by following a *business growth strategy*.
- Other entrepreneurs proactively thicken their agency through the combination of different activities (*diversification*).
- The reactive entrepreneurs do not thicken their agency, as they only conform to the structural conditions (*business preservation*).
- The entrepreneurs who follow an *education escape strategy* do not *want* to thicken their agency within the realm of their self-employed activity, as in reality they thrive for a high-prestige career job.

Specified research question regarding the recommendations for training providers and involved stakeholders
The final question looks beyond the actual evaluation and considers the outcome of the research endeavour, as well as the recommendations for the providers of entrepreneurship training regarding the improvement of structural conditions for disadvantaged youth. In this way, the research project comes full circle by bringing the focus back on the role of entrepreneurship training.

- **SQ8: What recommendations can be derived for the providers of entrepreneurship training and the involved stakeholders with regard to the improvement of structural conditions for young entrepreneurs from socially deprived back-grounds?**

4. Methodology

This chapter addresses the methodology of this dissertation. The research method needs to be in harmony with the object of research. From an epistemological point of view, the present research starts from the assumption that society is socially constructed through the activities of individuals. This applies to the construction of space as well (Mattissek et al. 2013 and Werlen 1997 from the perspective of Social Geography). Therefore, the aim of this dissertation's methodology is not to grasp the objective reality, but rather to reveal social processes. This places the individual as an actor at the centre of the study. Attention is focussed on two different types of actors, firstly on the providers of entrepreneurship training who are concerned with the empowerment of rural people, and secondly on the training graduates, young entrepreneurs who are seeking to build up sustainable livelihoods under certain structural conditions. From these actors, not only facts about their activities and social conditions are to be captured, but also their personal perceptions, convictions and motivations. Therefore, a *qualitative* research method is most feasible for both the collection and the analysis of the data. For the survey, guided interviews have been conducted with a manageable number of both representatives of training providers and entrepreneurs. For the analysis the method of the qualitative text analysis is chosen.

The qualitative analysis depends crucially on the research questions, which have been derived from the theoretical framework laid out in Chapter 3. After this, the survey instruments will be introduced, and thereafter each step of the qualitative text analysis with the specific method that is being used. This is followed by an explanation of the complete category system, including all subcategories and values. Finally, it will be necessary to make some important remarks regarding the quality criteria, including how to deal with language barriers.

4.1 The survey instruments

One of the strengths of qualitative research methods is that they can be adjusted to the research question and the specific topics that are derived from the theoretical framework, making the results more directly appropriate to the objective of research (Mayring 2016). For the inquiry of data, guided interviews seem the most appropriate instrument for evaluation, as they allow verbal access with the interviewees, giving them a voice as the subject of the research. As semi-structured interviews, they are problem-oriented (Witzel 1985), which means that they can be structured according to the specific topics given by the specified research questions. Guided interviews are open, insofar as they are not intended to provide all the answers, but allow the respondents to answer freely, putting them into a position where they must think for themselves. If new topics are brought up by the interviewee, ad hoc questions can be asked (Mayring 2016). If necessary, the interview guide can be further adjusted in the course of the qualitative analysis.

However, a different type of structure is chosen between the interviews with the entrepreneurs, who have completed an entrepreneurship training, and those with the representatives of training providers. For the entrepreneurs, the structure is somewhat tighter, as it is important to make them focussed on their specific topics of interest, and to compare their statements with each other easily. For the representatives of the training providers, a looser structure is chosen, in which the questions are posed more generally, as these interviewees are

T. Aberle, *Entrepreneurship Training in Rural Parts of Bihar/ India*, Perspektiven der Humangeographie, https://doi.org/10.1007/978-3-658-30008-1_4

experts and possess a wide knowledge of the structural conditions of the entrepreneurs, which needs to be captured. This supports the explorative character of the analysis. This having been said, the interview guide still needs to be semi-structured with regard to the main topics, so that the findings can be integrated into the category system and they can be triangulated with the findings of the entrepreneurs. Therefore, a narrative form of interviews (Schütze 1977) would have not been appropriate. But further experts have been included in the survey, e.g. a village head and village elder. For them, narrative interviews seemed appropriate, as only the general topics regarding the structural conditions of the respective villages need to be captured. The last group of interviewees are trainees who have been met within their teaching class and interviewed both in the group and individually. For this reason, semi-structured interviews were conducted. A few of the trainees have been chosen for individual interviews in order to obtain deeper insights on some of the topics, which are identical with the topics used for the interviews with the entrepreneurs. In the following section, the order of the topics of the respective interview guide for each group is introduced. The interview guides are found in the Appendices VI-IX.

Semi-structured interview guide for the entrepreneurs (Appendix VI)
The interview guide consists of five parts. The first part is a brief evaluation of relevant background information about the candidate, such as age, family background, education and qualification. The second part, the evaluation of the training course, is an opener to help the interviewee to focus on the training course and remember it. This part addresses the interviewee's motivation for the training, his or her ability to utilize specific training inputs, and whether the business could have been increased afterwards. It also asks the interviewee to grade the training course. The third part is the actual evaluation regarding the question of how access to the five capital assets could be increased. The interview guide also provides some variables which help to itemize each of these capital assets. The fourth part relates to the structural conditions and asks about structural limitations as well as the influence of socialization agents, which include the extent of family support in the business and benefits through handholding with the training provider. The final part focusses on the interviewee's personal, educational, and professional aspirations.

Semi-structured interview guide for expert interviews with representatives of the training providers (Appendix VII)
The information about the structural conditions regarding the empowerment of disadvantaged youth through entrepreneurship training is to a great extent gathered through expert interviews. The largest group among them are the representatives of the training providers. For them the interview guide is looser and more semi-structured. It contains general questions regarding the training institution and the offered training courses, and more specific questions regarding the handholding with the graduates and the different kinds of connections with the training provider, e.g. with the village community, banks or the business world.

Guidelines for the narrative interviews with village heads/elders (Appendix VIII)
The guidelines for the narrative interviews with village elders or heads only list topics that need

to be covered. These include socio-economical information about the village members, information regarding the village infrastructure, including educational institutions such as schools and training providers, and information regarding development initiatives of the government, NGOs, community-based organizations and the like within the village.

Semi-structured interviews with trainees in a group[54] and individually (Appendix IX)
The last group of people are trainees, which means that they have been interviewed while taking their entrepreneurship training. They have been included in order to verify some of the findings from the interviews with the entrepreneurs. The topics of interest include information regarding the interviewee's social background, educational status, their motivation for undertaking the training, and their personal, educational and professional aspirations. All these questions are also covered by the entrepreneurs, but due to time constraints the number of interviews with them had to be very limited, whereas interviews with trainees were shorter and a much greater number could have been included. Therefore, the question at hand is whether certain trends that have been identified by the entrepreneurs are also confirmed by the trainees, such as the fact that a great number of entrepreneurs have higher educational degrees, some of whom even come from lower sections of the society, and the fact that they often consider the entrepreneurship training only as a Plan B and have very ambitious career aspirations.

The same interview guide is used for both the group and individuals. Hence, the decision whether to ask a specific topic to the whole group or only to ask it to individuals was taken on the spot, depending on the feasibility at the time.

4.2 Description of the evaluated text material

This section concerns the important step between the acquisition of data through guided interviews and the text analysis. It introduces the procedure of bringing the original verbal information into a written form. Besides an audio device, a camera was also used during the visits of the interviewees to capture them visually. Its use and purpose will be explicated below. Since the survey has been conducted in rural parts of Bihar, an area where the average person is not able to communicate in English, a local translator was required to be present from the respective target regions during the interview session. Therefore, a section on bridging the language barrier has been included (4.7).

The transcription
All interviews have been recorded and then transcribed with the help of the program f4 Transkript. The transcribed version was then directly copied into MAXQDA for the text analysis, during which process the timings of the transcription were not carried over, as this was only relevant during the transcription of the recorded material. For the transcription, a "literal transcription" has been used with the goal of having a complete transcript of the text

---

[54] Here, the term "group discussion" is deliberately not used, since the group sessions consisted solely of a guided interview, where the answers had been collected from the whole group. The purpose of these sessions was to capture certain information from the largest possible number of respondents, but not to capture collective attitudes towards certain issues, which can be evoked through the group dynamic, for which the "Frankfurter Institut für Sozialforschung" developed the group discussion (Mangold 1960).

material (Mayring 2016, 89). This is required for the qualitative text analysis. But this decision has been taken with one restriction: the literal transcription has not been made in the original language, which usually was Hindi, but in English, which means that anything the translator said had been transcribed. For this reason, grammatical mistakes made by the translator have been corrected, as long as they did not change the original meaning. If the translator quoted an interviewee in the third person (e.g. "She is afraid of taking the risk") it was also transcribed in the third person, but it was transferred into the first person ("I am afraid of taking the risk") if it was used as a quotation for the analysis. Further aspects regarding the translation into English are considered under the section below on "bridging the language barrier".

A conversion of the text material into the international phonetic alphabet (IPA), which is a very accurate method for the literal transcription (Richter 1973), has not been considered as necessary, since no specific dialects or nuances had to be captured, but more importantly for our purposes the contents and themes.

One further question to be considered is the extent to which it makes sense to capture important information that goes beyond the literal transcription, such as emotions or pauses, which can be captured in the form of a commentated transcription, for which a practical system has been developed by Kallmeyer and Schütze (1976). This system has not been used here, since the primary interest lay upon the provided information and to a lesser degree on the way that this information was delivered. Only a few occurrences had been captured, for which the following individual notes have been used:

- (Pause): If the interviewee paused for a while before giving an answer.
- (Trans): If the translator did not translate, but made a comment or added something on his own.
- (Hindi speaking): Whenever the translator chatted with the interviewee in Hindi. Sometimes more family members had been present in the room, at times creating a small discussion among them. If that seemed relevant, the translator was approached afterwards for clarification.
- (?): If there was anything that the translator said which could not have been understood. In such cases, the translator was approached afterwards for clarification (see under "bridging the language barrier").
- (own remarks: …): sometimes, for the sake of clarification, the author's own remarks were added.
- Various actions during the interview session have been put in brackets, e.g. (interviewee points to a men) by saying "he is my father", in an interview with a handloom worker: (Interviewee shows his products and demonstrates working on the machine).

The use of a camera

For the sake of illustration, a digital camera was always brought along to be able to take pictures. Pictures were primarily taken from places of interest that related to the entrepreneurial activity, such as a Rangoli shop or beauty parlour, cows, different crops, class rooms, different tools, e.g. a charkha machine or a handloom. In addition, various impressions from the village environment have been captured. Some pictures have been taken of the interviewees, but only for personal use, in order to better remember the interview situation.

4.3 The process of the qualitative text analysis

For the field survey, the method of the *qualitative text analysis* is intentionally chosen, in order that the findings and insights should be directly taken from the text material provided by guided interviews with both the training providers and their graduates (Kuckartz 2014/2016, Mayring 2015). Kuckartz formulated three basic methods of the *qualitative text analysis*. In any given research topic, both the method of the *evaluative qualitative text analysis* and the method of the *type-building analysis* are most adequate in order to acquire relevant and meaningful results in the end. The first of these two is chosen because the focus lies on the empowerment of the disadvantaged youth. Empowerment is a *process*, and therefore this variable must be itemized in different quantifiable dimensions, such as personal, sociocultural or financial empowerment, which can be expressed with regard to different degrees of access to various capital assets, or to different levels of participation. However, the goal is not simply to identify different levels of empowerment, but to understand different coping strategies that are being used in order to deal with structural limitations. Therefore, a type-building analysis builds on the findings of the evaluative analysis in order to characterize what types of youth (based on gender, education, occupational group and social background) have chosen what types of livelihood strategy. This multi-stage process of categorizing, coding and presenting can be divided into 15 steps, within which steps one to nine are part of the evaluative qualitative text analysis, and steps ten to 15 are part of the type-building analysis.

4.3.1 Evaluative qualitative text analysis

The method of the "evaluative qualitative text analysis" is based on Kuckartz (2016, 123ff). The starting point is the general research question. During this part of the analysis a category system is created consisting of evaluative categories, which have been derived both deductively from the theoretical framework and inductively from the evaluated text material. The goal is to analyse the data material with the help of these categories, for which the first six specified research questions and associated hypotheses provide useful orientation. The detailed steps are illustrated in Figure 6a. Since the data triangulation (step 8) is also performed on the basis of category, it appeared justified for the analysis (Chapter 6) not to deal with steps 7 and 8 successively, but to include step 8 – depending on the respective category – within the category-based analysis (step 7) at the appropriate place. Step 9 is a binding element between the evaluative and the type-building analyses, which is an evaluation of the development of personal initiative across categories.

Figure 6a: The process of the evaluative qualitative text analysis

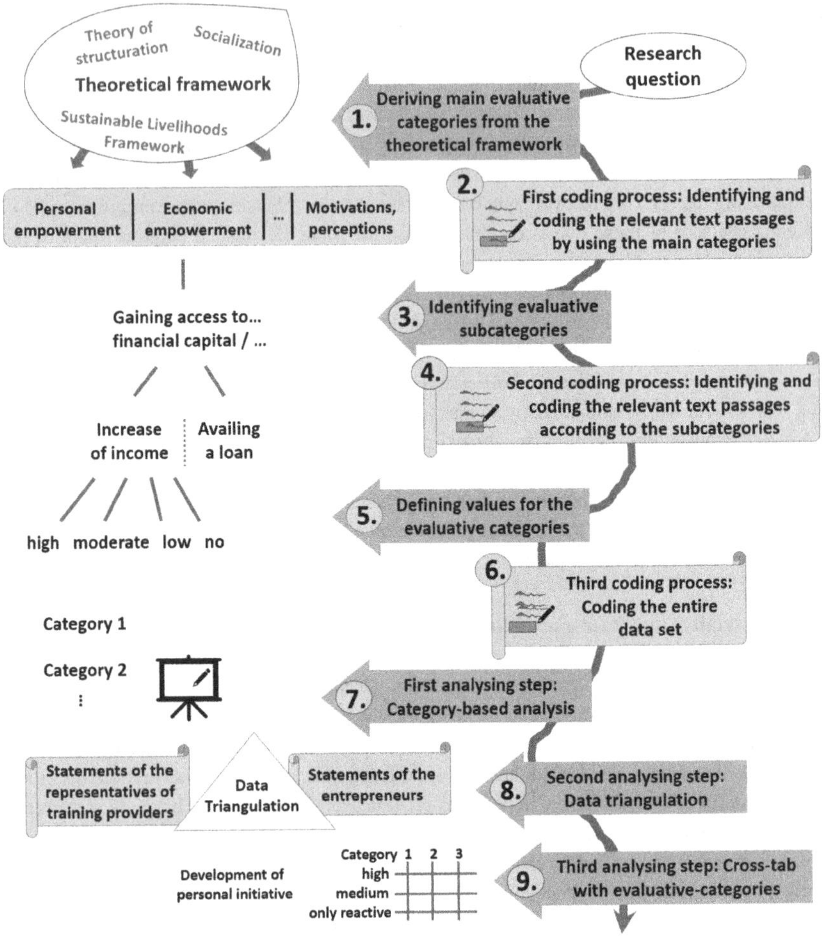

Source: Author's own draft, based on Kuckartz (2016).

4.3.2 Type-building analysis

The type-building analysis, which is based on Kuckartz (2016, 143ff), builds on the findings of the evaluative text analysis. The goal is to create "polythetic" types, which means types with heterogeneous attributes that have been inductively created from the text material. The basic principle is that "the elements of each type should be as similar as possible and the different types should be as un-like and heterogeneous as possible" (Kuckartz 2014, 105). The detailed steps are illustrated in Figure 6b.

Figure 6b: The process of the type-building text analysis

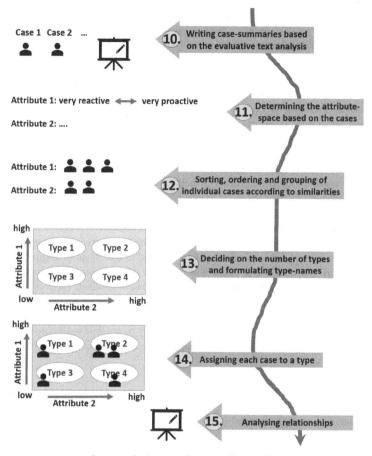

Source: Author's own draft, based on Kuckartz 2016.

The typology is based on attributes or characteristics. An example of a relevant attribute is the family background of a person, and an example of a relevant characteristic is a person's personal aspiration regarding career choice. Each of the entrepreneurs are then assigned to a type. In this way, the specified research question seven as well as its associated hypothesis provides a useful orientation to this process of categorization.

At the end of the analysis, the different livelihood strategies, which the entrepreneurs developed while coping with the structural conditions, will be visible. From this, conclusions regarding the role and the effectiveness of the entrepreneurship training during the process of empowerment can be drawn. This will enable us to reveal underlying issues and shortcomings that are rooted in the educational and societal system. This in turn provides – in accordance with the specified research question eight – a basis from which to point out suggestions for different stakeholders, such as training providers, community-based organizations and government institutions in working towards a more inclusive education system which opens up realistic perspectives for empowering disadvantaged youth.

4.4 The category system

A common category system has been created for both the entrepreneurs and the trainees, although some categories are only relevant to the entrepreneurs. The categories that apply to both groups are marked with "(+tr)", which means "also for trainees". A separate category system is used for the representatives of training providers and the village elders/heads.

Categories for entrepreneurs and trainees

The category system for the entrepreneurs has been constructed in accordance with action- and structuration-theoretical considerations (Giddens 1984). Due to the fact that every action is embedded within a structure, it is difficult to accurately group the categories into those which reflect agency and those which reflect structure. Yet it is reasonable to reflect in the category system the grouping illustrated in Figure 5 (Chapter 3) with the three circles denoting respectively agency, structure, and motivations and perceptions, so that it can be aligned with the specified research questions. These circles also reflect the three dimensions of empowerment according to Roy and Saini (2009). Thus, it seems appropriate to subdivide the category system into the following four parts (Figure 7):

1. personal empowerment, which entails the necessary background information regarding the accomplishment of the development tasks binding and qualifying (Hurrelmann 2012);
2. economic empowerment, reflecting agency;
3. socio-cultural empowerment, reflecting the structural set-up taken in consideration with the impact of socialization agents; and finally
4. "motivations and perceptions".

A definition as well as the respective values are given for each category. The division of the main categories into sub- or sub-sub-categories is marked by the use of different colour boxes. Whenever appropriate, quotations from the text material have been included to explain the reasoned choice of each category.

Categories for representatives of training providers and village elders/heads

In order to capture the insights from the representatives of training providers as well as the experts of the village community on relevant topics covered in the category system for the entrepreneurs, a triangulation of data is used. This means that each category for these expert groups has a comparison-category within the category system of the entrepreneurs, in which the following cross-reference is made at the due place:

> **Data triangulation with representatives of the training providers**

> **Data triangulation with village elder/head**

Hence, no separate category system is created for these expert groups of people. Additional information, that needs to be captured through these expert interviews, can be found in the Appendices VII and VIII.

A special note concerning the insights from the representatives of training providers regarding the entrepreneurs who did not establish a business successfully (see 6.1.2.3) should be made here. Specifically, the respective cross-reference in the category system for the entrepreneurs is made under the following sequence:

↳   Economic empowerment

↳   Business information

↳   Status of business (active/inactive).

The MAXQDA-Analysis
In order to gain deeper insights into the qualitative text analysis, the complete MAXQDA-Analysis can be requested from the author (to_aberle@yahoo.de), containing all interview scripts, the category system, and its respective codings.

Figure 7: The category system for the entrepreneurs and trainees

Part I: Personal empowerment

This background information helps to classify the entrepreneurs according to their youth status and their accomplishment of the developmental tasks "binding" and "qualifying" (Hurrelmann 2012). This gives a clue regarding their level of personal empowerment. Information regarding the concerned entrepreneurship training as well as further acquired training has been included as part of the personal empowerment.

Category (and subcategories) and definitions	Values
Gender (+tr)	Male/ female
Age (+tr) Young people between 18 and 35 are considered as "youth in transition", if they are still in education and/or unmarried (see 2.2.1).	18-19 20-25 26-35 above 35
Marital status (+tr)	Married/unmarried/unknown
Number of children (+tr)	No children; 1; 2-3; 4-5; above 5
Qualifying (+tr) School leaving qualification/university degree, with information whether the school or study course has been completed	BA or above 5th; 8th; 10th; 12th degree Did not attend school
Training info (+tr)	
Training provider (training of business)	Drishtee/ RSETI/ Fateh Help Society/ other
Training subject (training of business)	Dairy farming/ vermicomposting/ inland fisheries/ beauty parlour management/ Rangoli shop management/ stitching/ handloom/ charkha
Length of training (training of business)	Month/ year
Completion of training in...	Less than a month/ 1-3/ more than 3 month
Further training (including training provider/ length/ year)	
Related to: business of training/ further businesses/ not related to any business. Informal training/ taught at home/ no previous training or knowledge	

Part II: Economic empowerment

This category reflects the aspect of agency in the entrepreneurial activity, and therefore is subsumed under the term economic empowerment. It has three sets of subcategories:

- The business information, which contains important background information regarding the business in which the entrepreneurship training has been taken;
- Gaining or increasing access to financial, natural and physical capital; and
- The application of training inputs and development of personal initiative.

Whereas the first two sets of categories are basically an inquiry of standardized information, the last set provides qualitative information which reflect the specified research questions 1 and 2. Therefore, it appeared necessary only to support the last set with quotations from the text material.

Category (and subcategories) and definitions	Values

Business information	
Business sector (in which the training has been taken)	Dairy farming/ vermicomposting/ inland fisheries/ beauty parlour management/ Rangoli shop management/ stitching/ handloom/ charkha
Status of business	Active/ inactive (reason)
➢ **Data triangulation with training representatives**	
Additional businesses/responsibilities (in present and in the past)	Business sector employed/ wage worker/ housewife Did nothing after finishing or quitting school/ studies

Part II: Economic empowerment (continued)

Financial capital	
↳ **Income** (monthly in Indian Rupees)	above Rs 10 000; Rs 2 000 – 10 000; below Rs 2 000 Business is only self-supporting
Increase of finances (since training) Only if business was already active at the time of training	Income has at least doubled Increased at least by 1/3 Increased less than 1/3 No increase
Receiving a loan ➢ **Data triangulation with training representatives**	Loan directly from a bank Loan through farmers' club/ SHG No loan received Not applied for a loan Wants to apply for a loan
Availing other financial support ➢ **Data triangulation with training representatives**	Start-up capital from Drishtee Money through family/relatives/friends other (Government scheme/cooperatives)
Natural/physical capital	
	Equipment provided by training provider Purchases of land/cows/buffalo/tools/other since training No increase/decrease of natural-/physical capital Lack of natural-/physical capital (e.g. has no computer)

Category (and subcategories) and definitions	Values

Part II: Economic empowerment (continued)

These are evaluative qualitative categories, which are directly connected with the specified research questions 1 and 2.

Applying training inputs

This key category elaborates learning success through the training. Some interviewees only recapitulated things they had learned, whereas others explicated what they successfully could apply to the business. The differentiation between general/specific knowledge is made on the basis of the assumption that the more specific is the knowledge they can recapitulate, the greater the proportion of their discursive consciousness. This itself is assumed in accordance with the hypothesis for SQ1, which states that the more knowledge the entrepreneurs can transfer from practical consciousness to discursive consciousness through the training, the more effectively they can react with regard to their limiting structural conditions. The distinction between the acquisition of technical knowledge and entrepreneurial knowledge/competencies (see 2.2.3.1) appeared only to be relevant for the "specific things learnt/applied". Yet, in that category, only technical knowledge had been mentioned. The general statements in the categories "general things learnt/applied" simply referred to the overall management of the business. Therefore, no further specification had been made.

Quotations from the text material:

"(The training) enlarged my knowledge about how to perform day to day duties in rearing the cattle."	General things learnt
"I learnt which type of food is beneficial for the cow to produce more milk."	Specific things learnt
"From training I was able to manage the products properly, so how to behave with the customers."	General things applied
"I managed ... how much food to give to cattle, how to control diseases and how to clean the environment where cattle situated, so these are the benefits I took from the training."	Specific things applied
"...that much I knew earlier ... there was nothing new (that was taught in the training)."	Did not learn anything

Part II: Economic empowerment (continued)

Personal initiative
Any proactive behaviour to run the business better, to engage socially, to avail support (e.g. by acting beyond routine, tradition or what has been taught in training and without having a specific obligation to do something, acting proactively when facing problems, being determined to solve problems, having a long-term perspective). Even though the category of "development of personal initiative" will be considered after the category-based analysis in step 9 of the analysis, it seems appropriate to place it – in alignment with SQ1 and 2 – together with the "application of training inputs".

"I also started a beauty parlour in my Rangoli shop where I spend all the capital on my own."	Very proactive
"If any cow produces less milk, I will sell it and buy another one which produces more milk."	Some proactive behaviour
"I don't want to join (a SHG) ... because I need family, have no time to go outside the village for a meeting..."	Being solely reactive

Part III: Socio-cultural empowerment (impact of socialization agents)

These qualitative evaluative categories cover the impact of primary, secondary and tertiary socialization agents on the entrepreneurial activity and consider them as part of the structural set-up. They reflect the specified research questions 3, 4 and 5.

Family as primary socialization agent (+tr)
➤ **Data triangulation with training representatives**

This category considers the impact of the family on the entrepreneurial activity (SQ 3) and contains two sets:
- Family background (APL/BPL-status, number of siblings and household member, place);
- Important information on income-generating activities within the family, especially whether the entrepreneurial activity in which the training has been undertaken is a family business, as well as any other information regarding any further income-generating activities of the parents or other household members.

Family status (above/below poverty line) **(+tr)**

Families usually know their APL/BPL-status because of their APL- or BPL-ration cards, which they can obtain in order to get hold of social securities. "APL+" is not an official category, but some interviewees described themselves in this way. Due to large differences with respect to wealth this category is used in order to differentiate very rich people who belong to upper caste groups, have a lot of financial capital and own large properties, from average APL people.

BPL
APL
APL+

Number of siblings (+tr)

No siblings; 1; 2-3; 4-5; 6 or more

Number of household members (+tr)

1-6; more

Place info (+tr)

Village name; place of home/ moved

Part III: Socio-cultural empowerment (Impact of socialization agents continued)

Family as primary socialization agent (+tr) (continued)	
↳ **Family support for the entrepreneurial activity**	
"All the family supports me." – a Rangoli shop-keeper	Family business
"I have one sister, we both are running this shop."	Individual family members engage in business
"Is anyone else in your family supporting you in the stitching business? – No, I am doing it alone." alone…"	No family support

Further income-generating activities of family/household members (+tr)	
	Agricultural/ non-agricultural business Wage employment Business inactive/ retired

Handholding with the training provider

> **Data triangulation with the training provider**

This includes any form of handholding between the training provider and their graduates, which is targeted at supporting the entrepreneur to develop their business (see SQ 4).

"Any sort of help is given by that institution (RSETI)."	Gets support/advice
"Drishtee gives slivers (cotton) to me and when I process it then Drishtee buys it and I get 200 Rupees per Kg." – a charkha worker	Only stays in touch when it is necessary
"Drishtee has done some advertising at the initial stage (of that shop)."	Support has stopped
"Nobody was here to help me after the training." – a dairy farmer	No support
"I am in need of the support of Drishtee foundation because the income is not too much."	Wishes to get support

Category (and subcategories) and definitions   Values

Part III: Socio-cultural empowerment (Impact of socialization agents continued)

Engagement in social networks ➤ **Data triangulation with training representatives and village elders/heads** This includes any engagement in a community-based organization (such as a farmers' club, SHG or cooperative) or a political party/NGO, including informal types of support that a person gives or receives from others. Technically, these social networks could have been considered under "economic empowerment", as they primarily exist to gain access to finances. Nevertheless, they have a socio-cultural dimension, as they are an instrument to empower socially deprived groups in India (see 2.1.5 and Roy and Saini 2009,5). Hence, they are considered under the category of socio-cultural empowerment.	
	Engagement in... SHG/ farmers' club:
"Only one member from my family is part of a SHG" "I am the manager of Atma (a dairy farmers' club)."	Moderate engagement High engagement
	NGO/ political party:
"I am a member of BJP (a political party)." "I am in politics and I was the village head-man for two terms."	Moderate engagement High engagement
"I am also helping small farmers, buying sugar canes for them and sending it to the sugar mill."	Trains, mobilizes, and supports others
"I am still studying, so I don't want to engage in a SHG."	Not engaged in social networks

Part IV: Personal motivations and perceptions

This category includes any personal motivations, perceptions and aspirations, including one's awareness of structural limitations, one's motivation for taking the training and running the business, and one's future aspirations regarding education and work (see SQ 6).

Personal aspiration regarding ... (+tr)

L **Education (+tr)**

Aspirations regarding educational programs in which they are not yet enrolled (e.g. if someone in 12th grade is planning to study, or if a BA-student is considering doing a MA).

"I want to learn more about the economy and to know how to improve my own decisions for my animals." – a dairy farming trainee	Wants to connect education with self-employed activity
"If I get opportunity (I would go for more education)."	Generally open to pursue more education
"I need a BA of education (to become a school teacher)."	Specific plans to pursue a BA/MA/PhD
"Are you planning to go for more education? – No, now my children will get education."	Not interested in further education

Training (+tr)

"Do you want to take more skills training? – Yes, dairy farming...to get more knowledge."	Generally open to pursue further training
"I want to take training in beautification (to open a beauty parlour)."	Specific plans to take more training
"I am not interested in any more training...experience is giving me training."	Not interested in further training

84

Category (and subcategories) and definitions	Values

Part IV: Personal motivations and perceptions (continued)

Personal aspiration regarding … (+tr)

The entrepreneurial activity (+tr)

"I want to develop my business… but I don't have all the facilities available."	Wishes to increase business
"I want to increase my shop in spite of land or products – Do you also want to open a beauty parlour? – Yes."	Specific plans to increase business
"We don't want to explore this business more but after getting a BA-degree she (his daughter) wants to apply for a government job" – a father of a dairy entrepreneur	Is considering quitting the business

Apart from the entrepreneurial activity (+tr)

"I want to do any type of business where I get money."	Willing to do any business or to apply for any job
"If I get an opportunity to work with governmental organization then I would go for it."	Wants to get a government job
"School teacher from class 1 to 8."	Pursues another qualified job

Recruitment for training (+tr)

This category gives some tips for the impact of socialization agents. When asked the question "What motivated you to do the training?", many did not give any motives, but simply mentioned who recruited them for training.	Motivated through training provider (mobilizer)
	Motivated by family/neighbour
	Motivated by bank
	Personal motivation

Category (and subcategories) and definitions	Values

Part IV: Personal motivations and perceptions (continued)

Motivation for training (+tr)

"To become self-employed and to increase my income."	Becoming self-employed
"Through this training I wanted to increase my cattle's milk product, so that I can raise the income and also nourish my family."	Increasing self-employed activity
"Actually, I was selected by the bank for the loan and bank sent me to RSETI."	Getting credits from a bank
"I wanted the certificate so I can apply for a job."	Getting a paid job

Motivation for enrolling in/completing university studies (+tr)

E.g., if they are already enrolled at University or if they have completed their studies.

"I am planning to do a BA in Psychology ... to apply for a government job." – a stitching trainee	Specific professional goal
"Why do you go to university? – I want to become highly qualified as a very known person for my family and the society."	No specific goal

Training evaluation (Appraisal of the training from A to D)

This category helps the interviewee to bring what he or she has learned from the entrepreneurship training into the area of discursive consciousness and to reflect on it. The gradation from A to D helps to ensure this process of reflection. The respondents were also asked to give reasons for their appraisal, which is captured under praise/criticisms.	A (very good) B (good) C (acceptable) D (insufficient) Praise Criticisms

Category (and subcategories) and definitions	Values

Part IV: Personal motivations and perceptions (continued)

Awareness of limiting or enabling structures

This category not only reveals different areas of limitations but also helps to gauge the interviewee's level of awareness and their ability to reflect on enabling or limiting structural conditions of the entrepreneurs.

"Is the dairy farming business right now sufficient for you, like, do you get enough money to survive? – No"	Limitations regarding the development of the business
"Do you also want to go for more education, to get the 10th or even 12th degree? – Due to my many children I am unable to. When they are grown up I will go for more education."	Limitations regarding the achievement of educational goals
"Those people (the training providers) give them (small farmers) training to fill the papers, but after that they forget about them, they don't help them."	General comments regarding structural conditions
"Small farmers should become organized in a group, ... that will help them in development."	Constructive thoughts on how to cope with structural limitations
"I don't face any problem." – a Rangoli shop-keeper	Not being aware of any limitations
"In dairy farming, there is no such thing (like a seasonal crop failure),... you can get money for the whole year."	Aware of enabling structures

Source: Author's own draft.

4.5 Operationalization of the hypotheses

The following categories help to operationalize the hypotheses, which have been formulated under 3.2 in connection with some of the specified research questions:

- Hypothesis for SQ1: This reflexive knowledge consists of both the technical/ entrepreneurial knowledge and the perception of one's enabling or limiting structural conditions, which are captured by *Applying training inputs / Awareness of structural limitations*. In addition, the application of these to the self-employed activity is captured by the *Access to financial, natural and physical capital,* and insights regarding the ability to cope with structural limitations are derived from part III of the category-system, the socio-cultural empowerment.
- The hypothesis for SQ2 is captured – in addition to the previous categories – through *Personal initiative.*

- The hypothesis for SQ7 is built on the previous insights. The category system did not include any separate category for the different livelihood strategies, which are discovered through the type-building analysis. Conclusions regarding the selected livelihood strategies can be drawn from the following categories: *Personal initiative / Personal aspirations / Livelihood portfolios* (which will be considered in Figure 18 "household portfolio" under 6.1.3.1 "Family as primary socialization agent" below).

4.6 Quality criteria

Unlike in quantitative research, where standardized methods and static quality criteria are used in qualitative research, the quality criteria have to be adjusted to the specific set-up of the analysis. Therefore, they must be considered individually and made to conform with the research method and research target (Flick 1987; Mayring 2016). In quantitative research, quality criteria are essentially based on issues of validity, which refers to the question whether I have obtained the data that I wanted, and issues of reliability, which refer to the question whether I have captured the research objective adequately (Friedrichs 1985). This poses the question of how these criteria may be applied to qualitative research (Mayer 2013; Mayring 2015/2016). According to Mayring (2016), reliability is reflected in stability, which includes the inter coder reliability, accuracy and reproducibility, all of which in turn refer to the question whether the same results would occur if the survey was repeated under the same conditions. The question regarding validity is – just as it is in quantitative research – relevant in qualitative research with respect to the output of the study, namely whether those aspects which it was wanted to measure were actually measured. For example, if I want to measure the development of personal initiative, is it reflected in the identified proactive and reactive behaviour patterns (Mayer 2013)? According to Kuckartz (2016) and Miles et al. (2014), validity can be measured by means of external and internal validity, in which internal validity is indispensable in order to ensure external validity (Kuckartz 2016). In this dichotomy external validity refers to the representativeness of the study. As for internal validity, Kuckartz (2016, 204ff) formulated a practical check-list for internal quality criteria, which is very valuable to consider for the present analysis, as it connects directly to the method of the qualitative text analysis according to Kuckartz, which was used as our methodology. In the following section, those criteria that seem relevant are considered under the question of how they can be practically applied to the present analysis. In order to take into account the procedural character of the qualitative research, it seems feasible to consider these quality criteria for the three phases of data evaluation, description and analysis (Mayring 2016, 143). To this end, further insights are drawn from Mayring (2016, 144ff), who formulated six general quality criteria for qualitative research (see also Becker and Geer 1979). Finally, final considerations are made regarding the criteria for external validity.

Quality criteria within the process of the data evaluation

- With respect to the process of the data evaluation within qualitative research, Kuckartz (2016, 204) refers to the following criteria in connection with internal validity: Has the data material been recorded, and has a post-script of the interview situation been generated? For the present analysis these criteria have been taken seriously (see under 4.2).

For any interview, a portable recording device has been used for the recording, and a digital camera has been used in order to capture the interview situation visually. In addition, hand-notes have been made.

- Mayring (2016, 143) also gives consideration to the authenticity of the person who is being interviewed. First of all, this is ensured by selecting the interviewees very precisely on the basis of their professional background. All of them were actively engaged in the area of entrepreneurship training – the objective of research for the present analysis – either by taking the training or by providing the training. Therefore, a high competence for giving worthwhile answers regarding the topics directly linked with entrepreneurship training can be assumed.

- This relates to a further point, which Mayring (2016) lists among the six general quality criteria: the importance of ensuring proximity to the object of research. This was achieved by visiting all of the interviewees directly within the target region at their homes or work-places.

Quality criteria within the process of the description of data:

- With respect to the description of data, Kuckartz (2016, 204) highlights the importance of a complete description of the data material in abidance to certain transcription roles. This process has been documented in detail under 4.2.

- Kuckartz further underlines the importance of the fact that what is written must reflect what was said. Since this is connected to bridging the language barriers, it will be discussed in detail under 4.7.

Quality criteria within the process of the analysis

- With respect to the analysis, Kuckartz (2016, 204-205) refers to the fact that the chosen method must be adequate for the research question, and that the category system and its codings must be consistent. This is what Mayring (2015, 125ff) addresses by his term *semantic validity*. These aspects are considered under 4.3. At this point, only a few aspects are emphasized in particular. For the transcription and coding of data, no additional help was recruited except for dealing with translation issues. Therefore, the criteria of the inter-coder reliability is of no concern, yet issues connected to the auditability, i.e. the plausibility of codes and causal assumptions, must be taken into account. As multiple coders have not been used in the present work, its plausibility is ensured by strictly using the approach of the argumentative interpretation (cf. Mayring 2016, 145), which draws its validation from the theoretical framework. This relates to the earlier mentioned issue of the adequacy of the general and specified research question with the theoretical framework. Furthermore, the repeated coding over the course of the 15 steps of the analysis ensures a solid process of coding, allowing one to derive reliable conclusions from the evaluated data.

- Kuckartz also asks whether contradictory examples should be considered or if only the technique of so-called *selective plausibility* should be applied, which is a term used by

Flick (1989 and 1999). It describes the method of only citing those passages from the evaluated text material that represent the typical aspects of day-to-day routine, in order to understand social action (Girtler 1984). This technique is applied to the present analysis. In order to ensure greater plausibility, extreme cases are especially highlighted and analysed, e.g. the cases of those who did not manage to start their business successfully. Contrasting cases are also considered through the means of the type-building analysis.

- Mayring (2016, 147, in reference to Klüver 1979) further highlights the principle of *communicative validation*, by which is meant the capacity to acquire additional information from a third party for the validation of the evaluated text material. This principle was utilized in the present work by presenting the evaluated text material afterwards to the Drishtee leadership team, in order to obtain feedback and to discuss the findings.
- A further quality criteria listed by Mayring (2016, 147, in reference to Jick 1983) is the triangulation of data, which is a way to increase the quality of research by combining multiple steps of the analysis. In the present research this is achieved by considering multiple perspectives on certain topics from both the entrepreneurs and further expert groups.

Criteria regarding external validity

- In considering external validity, the question at hand is the extent to which the findings from the survey are generalizable (Kuckartz 2016, 217ff). This signifies an area of tension, since whilst on the one hand, only single cases are considered – which provide insights that certainly cannot be generalized *per se* – on the other hand, common patterns can be identified, which had been considered in the identification of the thematic categories for the qualitative text analysis. This directly connects to the next point, namely consideration of the representativeness of the selected people.
- Finally: is the number of cases representative? The representativeness of a study certainly increases with an increased number of cases. Since the present analysis follows a qualitative method, the number of cases had to be limited to a manageable size. This manageable size has certainly been reached with 25 entrepreneurs plus interviews with trainees and additional experts from the field. Hence, the interviewees have intentionally been selected from different research locations with similar conditions, but in different geographical areas, so that the findings of identical groups of people from various locations can be compared. In addition, multiple occupational groups have been given consideration in order that common patterns can be identified across these groups. Furthermore, the insights of entrepreneurs are compared with insights from the trainees. Finally, the earlier mentioned validation through triangulation and communicative validation have both been used in order to ensure a higher representativeness.

4.7 Bridging the language barrier

The issue of bridging the language barrier is especially important in qualitative research, where the whole analysis is based on the meaning of what the respondents said during the interview (Kruse et al. 2012). This issue needs even more attention in the present dissertation, where most of the interviewees spoke in their mother tongue and a translator always had to be present both to convert the questions from English into their mother language and to convert their answers back into English. For this reason, not only does the language barrier have to be bridged, but also the cultural differences, which means that words may have different meanings and customs of use in different cultures (Cappai 2003). This implies that every expression must be interpreted in its own cultural context, which poses challenges for the translator to first understand the custom of how a word is used before making the translation (Renn 2005). Thus, every translation is a projection. In this light, Temple and Young (2004, 171) note that "the translator makes assumptions about meaning equivalence that make her an analyst and cultural broker as much as a translator." Larkin et al. (2007, 468) point to the fact that "the translator has the potential to influence research significantly by virtue of his or her attempt to convey meaning from a language and culture that might be unknown to the researcher". In addition to the problem of cultural difference is the problem that non-equivalent words may exist in the other language (Kapborg and Berterö 2002). The challenge at hand is thus to minimize these problems, which requires due consideration of how the language barrier can be practically bridged within the present survey. At first, requirements for the interpreter are pointed out, before practical issues which occurred in the course of the present research and the ways in which they were resolved are outlined.

Language problems can be for a great part solved through a careful selection of the interpreter. Kapborg and Berterö (2002), who pose the question of whether the use of an interpreter in qualitative interviews threatens validity, refer to the cultural arena, which, according to Rubin and Rubin (1995, 22), is "not defined by a single belief or rule, or by a handful of phrases unique to the group, but by a whole set of understandings that is widely shared within a group or subgroup". According to Kapborg and Berterö (2002) validity is strengthened if the interpreter is part of the cultural arena. This complies with Freed's (1988) notion that it would be ideal if the culture of the interviewee and the interpreter were the same. Kapborg and Berterö (2002) further add that the interpreter should not only have the required language skills but also be trained in the respective research field. This is especially relevant in Bihar, due to the great diversity of language and culture. Even within Bihar, multiple languages are spoken.[55] Kapborg and Berterö (2002) identified two points of potential loss of meaning through the use of an interpreter, which are also relevant for the present survey: firstly, when the interpreter translates the researcher's question into the local language, and secondly, when he translates the answer back into English. Regarding the first case, the interpreter may use – because of the language barrier – "simplified questions that were easy to answer" (Kapborg and Berterö 2002, 55). Regarding the second case, there is a risk that the expressed reality in the local language could differ from the expressed reality of the translator in English.

In light of these concerns, the two interpreters who were used during the field survey – one for the Madhubani district and one for the districts of Bhagalpur and Munger – fulfilled the following requirements. Firstly, it was ensured that they both came from the same "cultural arena", which ensured both familiarity with the local languages and dialects and a strong degree

---

[55] In Bihar, Hindi is a statutory provincial language, which is also used for educational purposes. A further widely used language within the surveyed districts is Maithili (Lewis et al. 2015).

of cultural sensitivity. Secondly, they had attained a higher educational degree. Thirdly, they were not alienated from the research field. And fourthly, they were not part of Drishtee, even though they were mediated by Drishtee. Thanks to these precautions, the autonomy and reduction of bias on the part of the translators were ensured. The interpreter used for the Madhubani district was a journalist older than 45 from that region, who also used to be engaged in setting up a farmers' club, while the interpreter for Bhagalpur/Munger grew up in Bhagalpur and was a university student older than 25, who had already completed his BA.

These requirements ensured a higher level of understanding of the questions and answers and prevented them from simplifying or deviating in content. Furthermore, all interviews were recorded and transcribed in English, which allowed one to approach the interpreter and let him hear the recording afterwards in case of any ambiguities. If this was no longer possible, a further native Hindi speaker, who was an internationally travelled English teacher, was involved for the review of the recordings. This ensured the avoidance of shortcomings, something which Berman and Tyyskä (2011, 181) point to in saying that not everything that is being said is translated or that "interpreters change meanings by omission, revision, and reduction of content" (cf. also Aranguri et al. 2006).

5 The Field Survey in Bihar, India

This chapter focusses on the field survey in the Indian state of Bihar, which was conducted during the winter of 2015/2016. Its goal is to introduce the different training centres visited in the surveyed region, and to provide an overview regarding the scope of the survey, listing the number and the different types of people that have been interviewed. Figure 8 provides an overview of the two target regions in Bihar, which have been chosen as the regions where Drishtee – the main cooperation partner for this project – operates its training centres. The total length of stay in Bihar was ten weeks. In addition, four weeks were spent at the Drishtee headquarters in Noida, near to India's capital city of New Delhi.

Target region 1 is the district of Madhubani, which is in the northern part of Bihar on the border with Nepal. Target region 2 comprises the two neighbouring districts of Bhagalpur and Munger, which both lie southwest of Bihar at the Ganges river. Figure 8 consists of two layers, one for the visited training locations and the other for the visited villages. In addition, it shows where all the different entrepreneurs and village heads or elders have been interviewed. The interviews with the training representatives and trainees were conducted at the respective training locations, and they are not included in Figure 8. Each entrepreneur has been labelled with an ID to ensure their exact identification in the further analysis. The order of the numbers complies firstly with the research location (Madhubani, Bhagalpur and Munger), and secondly with the order of the interview. The first letter stands for the occupation: D = Dairy farming; F = (Inland) Fishery; V = Vermicomposting; C = Charka (spinning); H = Handloom (weaving); S = Stitching; L = Ladies shop (Rangoli shop); and B = Beauty parlour. The second letter, "E" stands for "entrepreneur".[56]

In the district of Madhubani, only the visited training locations are shown. The visited entrepreneurs, who conducted their training at these locations, were all visited within their villages at their homes or work places. Yet, these villages are not shown on the map for the sake of clarity, as they were very scattered within a radius of approximately 20 kilometres around the village of Saurath. However, in the target region 2, the visited villages are all shown, as they exemplify the existence of a cluster for different occupational groups (see 5.1 for further information on this). In the area of textiles, the different steps of the production chain can be retraced (see further 5.2). If only a training centre and not the private homes were visited within a village, only the symbol of the training centre is shown on the map with the name of the village, e.g. Drishtee Khagra.

---

[56] For example: "DE2" is the second dairy entrepreneur, which has been interviewed at the first research location, namely the district of Madhubani in Bihar.

© The Editor(s) (if applicable) and The Author(s), under exclusive license to Springer Fachmedien Wiesbaden GmbH, part of Springer Nature 2020
T. Aberle, *Entrepreneurship Training in Rural Parts of Bihar/ India*, Perspektiven der Humangeographie, https://doi.org/10.1007/978-3-658-30008-1_5

Figure 8: The target regions in the state of Bihar, India

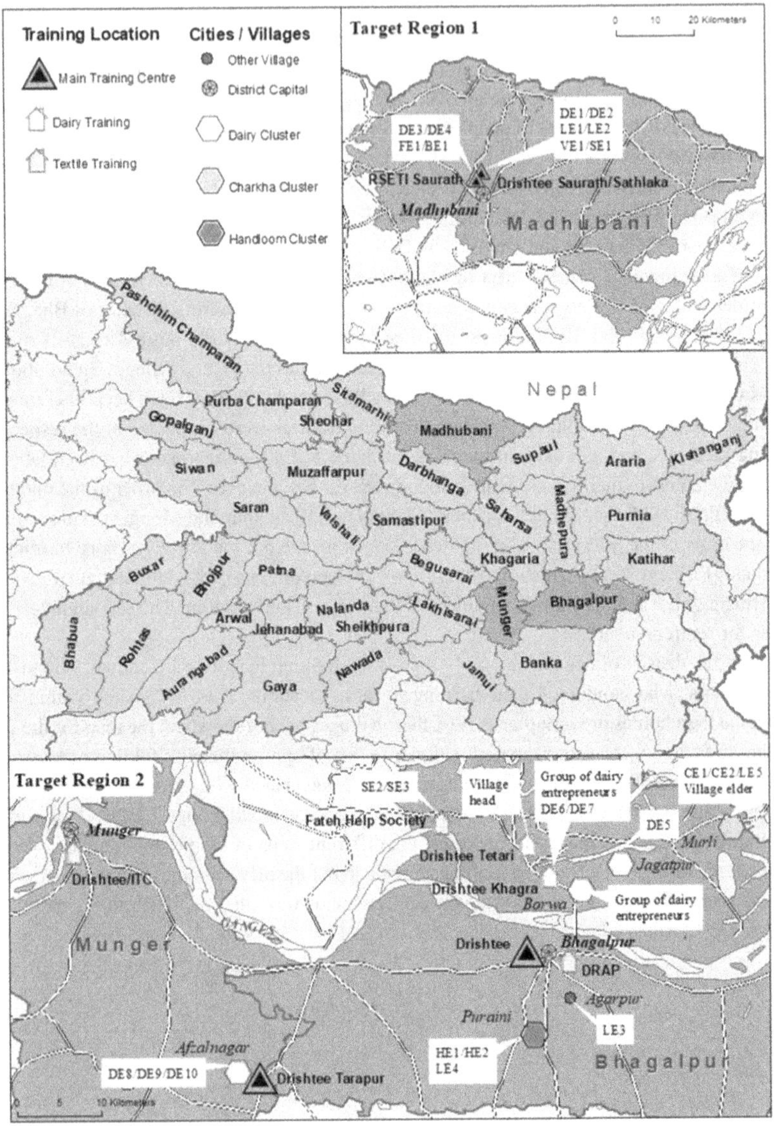

Source: Author's own draft, created with ArcView. Maps and its features powered by ESRI.

5.1 The surveyed entrepreneurs considered with their village environment

Most of the entrepreneurs – except the beautician and two dairy farmers, who were trained by RSETI in the Madhubani district – have been trained by Drishtee, or, as in the case of the handloom workers, are in a business relationship with Drishtee. Since Drishtee, as part of its business philosophy, is engaged with developing sustainable communities and thus follows a holistic development approach, in which it selected the villages very deliberately based on the skills and professions that were already available, it is helpful to know which entrepreneur belonged to which village and what each village is unique for. In the district of Bhagalpur, the charkha workers (CE1 and CE2) were met by the researcher in the village of Murli. In the same village, a village elder as well as a Rangoli shop-keeper (LE5) were also interviewed. The handloom (HE1 and HE2) workers were visited in the village of Puraini, which is part of the handloom cluster in Bhagalpur. There, the Rangoli shop-keeper LE4 was also visited, who came from a handloom family. The other three Rangoli shop-keepers were from different villages in Madhubani (LE1 and LE2) and Agarpur in Bhagalpur (LE3). Out of the dairy entrepreneurs, twelve were met in the village of Khagra within a group session (Bhagalpur district), one in Jagatpur (Bhagalpur), three in the village of Afzalnagar (Munger), and four in different villages in Madhubani. For both the charkha workers and the ladies' shop-keepers, it must be noted that the initiative to set up these businesses came from Drishtee as a part of its strategy to empower rural woman. The ladies who opened a "Rangoli shop" were provided with all the equipment and a credit of 20 000 Rupees by Drishtee, and continued to be supervised by Drishtee. Likewise, the charkha machines were given to the ladies by Drihstee to make yarn, which they again sell to Drishtee and which, in turn, will be further sold to the handloom workers. The Rangoli shop was in all cases run with the help of further family members. In the case of dairying and handloom, it is strongly connected with the family tradition, which means that they have learnt it from scratch and developed a routine, whereas the others had to familiarize themselves with it.

5.2 The training centres visited within the target-regions

The different providers of entrepreneurship training that were included in the present research have already been introduced in Chapter 2.[57] At this point, only an overview is given regarding the locations of each visited training centre, including the types of training that are being offered there.

Training centres visited in the district of Madhubani, Bihar

In the district of Madhubani two different training locations were visited, namely Drishtee and RSETI. The main Drishtee training centre in Madhubani is in the village of Saurath. Here, two types of training are offered. The first operates under Deendayal Upadhyaya Gramin Kaushalya Yojana (DDUGKY)[58], a skill development program which specifically addresses rural youth from poor backgrounds, and the second operates under PMKVY (see 2.2.3.4). The PMKVY-

---

[57] For Rural Self-Employment Trainings Institutes (RSETIs) see under 2.2.3.5, and for Drishtee and Fateh Help Society see under 2.2.3.6.

[58] This program is named after Deendayal Upadhyaya (1916-1968), one of the leaders of the Bharatiya Jana Sangh Party, which later on became the Bharatiya Janata Party (BJP).

training courses are run in the village of Sathlaka, where interviews were conducted with dairy farming and stitching trainees, as well as with representatives of Drishtee. The entrepreneurs who received their training through Drishtee were interviewed in the areas of dairy farming, vermicomposting and stitching. In addition, two female entrepreneurs were interviewed, who earlier received an entrepreneurship training at Drishtee and who are now operating so-called Rangoli shops, which are retail outlets in villages, serving woman with cosmetic products.

The RSETI in the Madhubani district, which has been included in the field survey, is also located in the village of Saurath. Interviews were conducted with four self-employed RSETI-graduates in the areas of dairy farming, inland fishery and beautification. According to personal information from the director, the RSETI-institution had aligned its training courses with PMKVY.

Training centres visited in the district of Bhagalpur, Bihar
As Bhagalpur is a cluster for textile production, it seemed appropriate to include a further private training institution that specialised in it, in order to obtain a broader view in the areas of textile production and women entrepreneurship. Therefore, the Fateh Help Society was visited in a village at the north side of the Ganges, where two ladies (SE2 and SE3) were interviewed, who took an entrepreneurship training in handbag-making. Due to different circumstances they were not able to continue this activity.

In Bhagalpur Drishtee supports rural entrepreneurs in the making of textiles and covers the whole chain of production, as is illustrated in Figure 9. During the first stage, yarn is being made by ladies who work from home on a self-employed basis with a charkha machine, which was provided to them by Drishtee. During the second stage, the yarn is delivered to traditional villages that are part of a handloom cluster. Here, weavers make clothing materials from it. In the last stage, Drishtee buys these materials from the weavers and processes them at the Drishtee Rural Apparel Production-Centre (DRAP), where sewers are employed to design garments from it. These clothes will finally be sold either at various market places in India or online.

Figure 9: The production chain in the making of textiles

Charkha workers   *Handloom worker* *DRAP-Centre*

Source: Author's own pictures.

In the area of textiles, additional interviews apart from the entrepreneurs were conducted with stitching trainees at the DRAP-centre. In the area of dairy farming, different training locations of Drishtee were visited in villages, which are partially offered on school properties.

There, additional interviews have been conducted with both training representatives and trainees.

Training centres visited in the district of Munger, Bihar
In the district of Munger, two training centres of Drishtee were visited. At both places interviews were conducted with stitching trainees. One of these locations is operated in partnership with the Indian Tobacco Company (ITC), where ladies from poor backgrounds, who were previously making incense sticks, had been retrained in stitching, as the factory in which they were employed was about to close down.

5.3 The scope of the survey
Next, the scope of the survey with regard to the categories of people and occupational groups will be introduced.

Categories of people
This survey captures first and foremost the entrepreneurs. However, additional groups of people had to be included in order to obtain a full understanding of the structural conditions in which they develop sustainable livelihoods. These are specifically the representatives of training providers, the trainees and further village experts. At the end of the survey, a presentation and discussion of the findings was conducted with members of the leadership team at the headquarters of Drishtee in Noida, which will be mentioned in further detail below.

Group 1: Entrepreneurs
The main target group are – in agreement with the title of this research project – disadvantaged youth from rural parts of Bihar, who have undergone an entrepreneurship training and are engaged in a self-employed activity. Ten of all the entrepreneurs fall under this category, six of whom are female, and four of whom are male.[59] However, in order to construct a realistic picture regarding the statements of the involved disadvantaged youth, additional entrepreneurs have been included as well. These are young people from above the poverty line as well as more mature entrepreneurs above the age of 35 from both below and above the poverty line. Out of the total number of 25 entrepreneurs are 14 young people (five male/nine female). A detailed breakdown of these subjects according to occupational groups is given below. In addition, two group-interview sessions were conducted with dairy entrepreneurs in the villages of Borwa and Khagra, Bhagalpur district, in order to gain insights regarding the access of these entrepreneurs to various livelihood assets as well as their personal motivations and perceptions whilst using a greater number of people. Normally, all the surveyed entrepreneurs had completed an entrepreneurship training at one of the visited training locations, yet there are the following exceptions. The two entrepreneurs in handloom had not done an entrepreneurship training with Drishtee, and two other entrepreneurs were currently in training at the time of the interview, but are considered together with the other entrepreneurs, as they had already established an

---

[59] One of these ten young people had stopped the entrepreneurial activity at the time of the interview.

active business.

All entrepreneurs except for DE6 and DE7, who were part of the Khagra-group, were visited at their place of work, which was usually also their home. Very often, the father of the entrepreneur or additional household members were present when the interview was being carried out.

Group 2: Representatives of training providers

This group has been included to capture information regarding the training courses and modalities, the handholding with the graduates, and to better understand their perspectives of becoming self-employed in their respective areas. The training managers had been interviewed at all visited locations. Wherever training classes had been visited to conduct interviews with the trainees, the teacher had also been interviewed. The types of interviews conducted involved the following personnel:

- **Training managers** (of various training locations). This could be the director of a training centre or a department leader within the training centre. Altogether, eight interviews were conducted.
- **Different subject teachers:** Interviews were conducted with ten teachers, five in dairy farming, and five in stitching.
- **Training mobilizers:** All the training institutions used mobilizers for the training recruitment. One interview with a mobilizer was conducted in the district of Bhagalpur.

Group 3: Trainees

These trainees consist of young people in the age-bracket of 18-35, who were attending an entrepreneurship training course under PMKVY. Group interviews were conducted in eight different training classes, four in dairy training and four in stitching training. Usually, individual trainees had been selected as well for separate interviews. The reason for not including further subjects, such as entrepreneurship training to open a Rangoli shop, is that the respective training was not offered at the time of the author's visit.

Group 4: Further experts of the village communities

Interviews were conducted with a village head and a village elder from two different villages in the district of Bhagalpur, where Drishtee is working with the goal of developing sustainable communities.

Group 5: Drishtee leadership in Noida, Uttar Pradesh

A presentation and discussion regarding the field survey and the first findings was conducted at the end of the survey with representatives of the management team of Drishtee at their headquarters in Noida. Among them was the co-founder and Managing Director of Drishtee, Satyan Mishra. This session was also recorded.

Table 6 summarizes the scope of the survey. It differentiates all conducted interviews by the target-district in Bihar, India, the group of people, and the occupational sector. It only lists the individual interviews. The group-interview sessions are not listed separately.

Table 6: Interviews with entrepreneurs, experts and trainees arranged by district and occupational group

District/occupation. group	Entrepreneurs	Experts/trainees
Madhubani district		2 managers (Drishtee)
		1 manager (RSETI)
Dairy farming	4 males (2 Drishtee/ 2 RESTI)	1 teacher / 2 trainees (Drishtee)
Inland fishery	1 female (RSETI)	
Vermicomposting	1 male (Drishtee)	
Stitching	1 female (Drishtee)	1 teacher / 2 trainees (Drishtee)
Ladies shop (Rangoli)	2 females (Drishtee)	
Beauti parlour	1 female (RSETI)	
Bhagalpur district		2 Drishtee managers (training and DRAP)
		1 manager (Fateh)
		1 mobilizer (Drishtee)
		2 village head/elder
Dairy farming	1 male / 2 females (Drishtee)	3 teachers / 4 trainees (Drishtee)
Charkha (spinning)	2 females (Drishtee)	
Handloom (weaving)	2 males (connected with Drishtee)	
Stitching	2 females (Fateh)	1 teacher / 2 trainees (Drishtee)
Ladies shop (Rangoli)	3 females (Drishtee)	
Munger district		2 managers (Drishtee)
Dairy farming	2 males / 1 female (Drishtee)	1 teacher / 2 trainees (Drishtee)
Stitching		3 teachers / 4 trainees (Drishtee)

Source: Author's own draft.

99

Summary of the number of individual interviews by the different groups of people

Interviews with entrepreneurs: 25

Interviews with managers: 8

Interviews with teachers: 10

Interviews with trainees: 16

Interviews with village heads/elders: 2

Interview with mobilizers: 1

Total number of interviews: 62

6. Analysis

This chapter presents the findings of the qualitative text analysis. This analysis follows the methodology introduced in Chapter 4, which itself aligns with the general and the specified research questions set out in Chapter 3. The first part of this chapter, the evaluative qualitative text analysis, starts with a category-based analysis of the findings from the entrepreneurs, which adheres to the category system, being divided up by the different dimensions of empowerment (6.1). It seems reasonable to intertwine the following analytical step, the data triangulation – which relates the findings from the entrepreneurs with insights from the representatives of the training providers – with the previous one, if and where appropriate. This is what has been done whenever a category used for the entrepreneurs has an equivalent category used for the representatives of the training providers. The goal of the data triangulation is to obtain deeper insights regarding the underlying enabling or constraining structural conditions, that have led to the increase, decrease or stagnation of the business. After this, 6.2 addresses the importance of the development of personal initiative, 6.3 the awareness of enabling or limiting structures, and 6.4 the compliance of the entrepreneurial activity with the educational and career aspirations of the entrepreneurs. The second part of this chapter, the type-building analysis (6.5-6.8), groups the entrepreneurs according to their different livelihood strategies, which they developed to gain or increase access to sustainable livelihoods.

Part I – The evaluative qualitative text analysis

6.1 The entrepreneurs – are they trained and empowered? A category-based analysis

Are the surveyed entrepreneurs, who have undergone an entrepreneurship training, empowered to cope with structural limitations and to build up sustainable livelihoods? This is the question that will be explored through the category-based analysis, which provides the basis for the successive exploration of SQ1 and SQs 3-5.[60] Firstly, the entrepreneurs are grouped according to their youth status and their accomplishment of the development tasks of binding and qualifying. This allows conclusions to be drawn regarding their personal empowerment. Thereafter, the evaluative categories regarding their economic empowerment are considered. In this way, light will be shed on the first specified research question, namely the extent to which the graduates have been able to utilize specific training inputs of the entrepreneurship training course. Next, the evaluative categories regarding the impact of primary, secondary and tertiary socialization agents are explored, from which insights regarding the socio-cultural empowerment of the entrepreneurs will be gained.

6.1.1 Personal empowerment

The dimension of personal empowerment or self-empowerment includes the ability to achieve personal goals in life (Roy and Saini 2009, 5), and will be considered in connection with the accomplishment of various developmental tasks (Hurrelmann 2012). Hence, this section starts

---

[60] The SQ2 will be evaluated below in section 6.2.

with a grouping of the entrepreneurs in accordance with the definition of youth provided under 2.2.1 above (see Figure 10). After this, their social background, especially whether they are located below or above the poverty line (APL/BPL), will be captured as a reference point to classify them as more or less privileged. Finally, their accomplishment of the developmental tasks of qualifying and binding will be discussed.

6.1.1.1 Grouping the entrepreneurs as "youth" and as adults

Figure 10 illustrates the grouping of the entrepreneurs in connection with their age, gender, qualification, marital status and APL/BPL-status. Each entrepreneur has been labelled with an ID, to ensure their exact identification in the further analysis. The first letter stands for the occupation, the second an ("E") for "entrepreneur", and the order of the numbers is determined firstly by the individual's research location (Madhubani, Bhagalpur and Munger) and secondly by his or her order in the interview.

In connection with the accomplishment of developmental tasks, there are six individuals listed as "youth in transition", who are below the age of 35 and who either did not complete their school or university education or are unmarried. Among them are two from a BPL-background. The remaining eight entrepreneurs below the age of 35[61] all come from a BPL-background. Thus, there are in total 14 young entrepreneurs from the three main occupational sectors, dairy farming, textile and beauty and wellness, and one entrepreneur who conducted her entrepreneurship training in inland fisheries. As an "inactive" entrepreneur, who did not begin with a self-employed activity, she was included in the study as a negative example in order to reveal structural constraints.

There are eleven adult entrepreneurs above the age of 35 who have been deliberately included as a comparison group. They come from the same occupational groups, except for one entrepreneur who conducted his entrepreneurship training in vermicomposting. Noticeable are two female entrepreneurs, who stopped the entrepreneurial activity. They have been trained in the making of hand-bags by the Fateh Help Society, but transitioned to other types of occupation. They too have been included as a negative example, in order to obtain further insights into the opportunities for empowerment through entrepreneurship training.

---

[61] For the remaining entrepreneurs below the age of 35, no specific term (e.g. "young adults") is used, as the "youth in transition" are also young adults. Compare 2.2.1 and footnote 18.

Figure 10: Grouping of the entrepreneurs

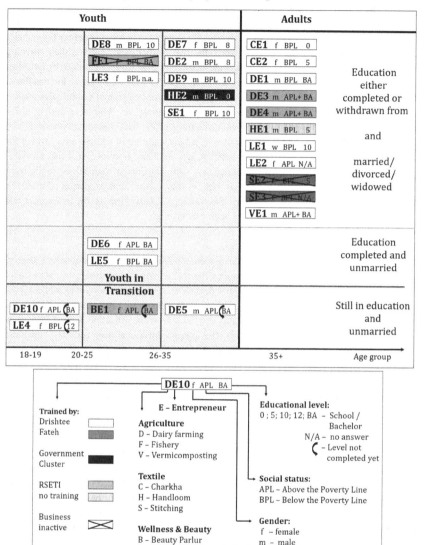

Source: Author's own draft.

6.1.1.2 The social status of the entrepreneurs

The entrepreneurs were intentionally chosen from both BPL- and APL-backgrounds in order to identify different degrees of structural limitations. The range varies from very poor to very rich. All the surveyed training providers were specifically targeted in order to empower the socially most deprived sections of the society. Hence, the two women who received their stitching training by the Fateh Help Society, as well as the charkha women who were trained by Drishtee, came from a very poor background. The handloom workers who were part of the handloom cluster in Bhagalpur, and the Rangoli shop-keepers who were trained by Drishtee, also came – with one exception – from a BPL-background. In contrast, the biggest group of dairy entrepreneurs was very diverse. Half of them came from a BPL-background, and some of them belonged to a village in the Munger district with a long tradition of cattle-rearing, which for them was closely connected with their caste. A few of them, however, still had an APL-background. The three APL-plus entrepreneurs, who considered themselves to be very progressive farmers, were classified as very rich, as they owned large properties of land. For them, the subject in which they received the training – dairy farming (in two cases) and vermicomposting – was only an additional source of income.

6.1.1.3 Qualifying as a developmental task

The level of qualification ranges from having no education to having a Bachelor's degree. Noticeable is the high educational level among the six "youths in transition", who were either pursuing or had already attained a Bachelor's degree, except for one very young student, who was still in Class twelve at school. Two of them are females from a BPL-background. Yet the positive correlation between the social status and the level of qualification is most evident by looking at the very poor and the very rich. The lowest educational achievements, which are demarcated by those who have between zero and five years of schooling, are held by the charkha workers as well as the handloom workers. In contrast, all people from an APL-background have either pursued or have already acquired a Bachelor's degree, except for the young entrepreneur who was still in school and one Rangoli shop-keeper, whose educational status was unknown.

It appeared that, for the youth in transition, especially for the very young ones aged 18-19, the accomplishment of the developmental task of qualifying had a significant impact on their capacity to focus on the entrepreneurial activity, and *vice versa*. In the case of the young Rangoli shop-keeper (LE4) who was still in the twelfth class, both her mother and sister helped out in the shop when she was at school. She mentioned that she would like to take a training course in beautification in order to advance the business, but she needed to postpone this plan due to her school commitments. In light of her social background – she came, like the two handloom workers with very low educational degrees, from a handloom family based in the village of Puraini in Bhagalpur, but from a different family – what stands out is the fact, that even though she had the privilege to attend a higher secondary school, she must have been only 16 or 17 years old when she took the entrepreneurship training to open the Rangoli shop, which took place over a year before the time of the interview, when she was 18 years old. In the case of the other young dairy entrepreneur (DE10) in the age group 18-19, who was pursuing a Bachelor's degree and who was at least 18 or 19 years old when she took the dairy training, her father explained that they were not planning to explore the business any further, as she wanted to apply for a government job once she received a Bachelor's degree. This leads to the conclusion that the more privileged woman from an APL-background was less distracted in achieving her educational goal than the one from the BPL-background.

6.1.1.4 Binding as a developmental task

The surveyed entrepreneurs confirm the trend – in contrast to mainstream Western cultures – that, especially in rural parts of India, both males and females often get married at a very early age. Among the surveyed young adults (aged 18 to 35), two in the age group 18-19 had not married, half of the six persons within the age group 20-25 had married, and in the age group 26-35 five had married (or divorced/widowed), with only one still remaining unmarried. In the age group of 35 and older, all entrepreneurs had married (or divorced/widowed). Taken in consideration with the accomplishment of the developmental task of qualifying, it stands out that all those who did not complete their school or university education were still unmarried, a datum which points to the fact that marital status hardly harmonizes positively with the continuation of the school or university education. This relationship is further complicated by the fact that, in all cases when they were married, they also had children with their married partners, which deters them even more from completing their education or achieving a higher educational degree. This was confirmed by a group interview session with twelve young female dairy entrepreneurs in the village of Khagra, Bhagalpur district. One lady from that group dropped out after the tenth grade due to her marriage. Having said this, it must be taken into account that females usually move to their spouse's family upon marriage, which determines their future engagement in a self-employed activity or professional career, and thus may deter some of the female entrepreneurs of the study from completing their higher educational degree and pursuing their own career aspirations.

Viewed in consideration with the personal empowerment of young entrepreneurs from rural parts of Bihar, it can be concluded that a high share of them attained a higher educational degree, yet a lower social status made it more difficult for them to pursue a professional career, as they were often compelled to engage in income-generating activities, which is usually the family business, from a very early stage. Women from poor backgrounds were even more restrained from pursuing higher educational goals, as they were often compelled to interrupt their education by the time they get married. In addition, among the young entrepreneurs who were pursuing or had already attained a Bachelor's degree, no one had combined their studies with the entrepreneurial activity, which opens up the question of how the entrepreneurial activity complies with their personal educational and professional goals.[62] This will be further elaborated under 6.4.

6.1.2 Economic empowerment

In this subsection a detailed look is taken at the economic empowerment connected with the entrepreneurial activity. Through a category-based analysis, light is shed on the question

---

[62] Out of the 14 young entrepreneurs, six were studying or had finished their studies (among them five females and only one male, half of which came from a BPL-background) in Sociology, Hindi, two times in History and two times in Political Sciences.

whether the surveyed entrepreneurs gained or increased access to financial, physical and natural capital assets after conducting the training.[63] At first, a sectoral analysis of the average income at the time of interview is given (Figure 11). After this, the increase of business since the individual's training is captured, which applies to those who had already established their business at the time of training. This category considers the increase of income since the training and – in the case of the dairy farmers – the number of cows or buffalo before/after the training, since these two are obviously interrelated (Figures 12 and 13). Next, the three entrepreneurs who did not manage to successfully build up a business are considered as negative examples in contrast to the other entrepreneurs. This comparison will be carried out by the means of a triangulation of insights from both the entrepreneurs and the representatives of the training providers. Thereafter, the individual entrepreneur's access to financial support, which means gaining access to credits or other sources of financial support, will be considered, also by means of a triangulation (Figure 14). As the intended goal of this section is to obtain insights into the specified research question, namely the extent to which the graduates have been able to utilize specific training inputs, these findings will finally be linked with the application of training inputs (Figures 15 and 16). Then they will be placed in connection with the entrepreneurs' evaluation of the training course. For the sake of comparability, a differentiation in the analysis is drawn between youth and non-youth and by the occupational sector.

6.1.2.1 Average income of the entrepreneurs

Figure 11 shows the average income, sector-by-sector, for the youth and the adults at the time of interview. Those who conducted an entrepreneurship training but did not successfully start the business of training are not included; this applies to the two women trained by the Fateh Help Society in the Bhagalpur district, and the young lady trained in inland fisheries by RSETI in the Madhubani district. These cases will be considered separately. The adult entrepreneurs (above the age of 35) who received their training in dairying or vermicomposting have also not been included at this point, as none of them provided information regarding their specific income through these activities. All of them were farmers from an APL-background with their own properties of land and diversified cultivations, while three of them could be considered as APL-plus, as their total income could be presumed to be 100 000 Rupees[64] and above, of which dairying is most likely only a small portion.

For several sectoral groups either the youth or adults have been interviewed. This applies to stitching and the beautician sector, where only the youth have been surveyed, and to charkha workers, where only the adults have been surveyed. The number of young dairy entrepreneurs is 17. This is because ten female entrepreneurs from a group interview session in Khagra[65] have been considered in addition to the seven young entrepreneurs that have been interviewed individually. This seemed reasonable due to the small sample size. In Khagra, twelve female entrepreneurs (eleven of them from APL) had been interviewed within a group session. They

---

[63] Access to social capital will be considered in 6.1.3 (socio-cultural empowerment).
[64] Or what is approximately 1 270 Euros.
[65] Two group interviews with dairy entrepreneurs had been conducted in two different villages of the Bhagalpur district, namely Borwah and Khagra. But in Borwah only information regarding the number of cows, and not the monthly income, had been captured.

all completed the same batch of the Drishtee dairy training. That dairy class, which had 21 women in it (twelve from APL and nine from BPL), had been visited while the training was still going on. The second visit, consisting of the group session with the twelve ladies, was conducted one month later, only ten days after the training was completed. Since two of these twelve ladies had also been interviewed individually, they are considered together with the other individual entrepreneurs, so the number of the entrepreneurs within the Khagra-group is only given as ten.

At this point an important preliminary note must be made. The given numbers regarding the income and its increase since the training do not claim to be absolutely correct, but can only be taken as estimates. This is due to the following reasons. When asked about their income, most entrepreneurs did not give accurate numbers, but answered a rounded figure to the full hundreds or thousands, while others gave a range, e.g. 500-600 Rupees. The given numbers have always been taken as it was said, and they have not been verified, e.g. by checking the individual's bank accounts or some similar means. If this had been done in every case, it would surpass the scope of the survey. Furthermore, the goal of the present survey is not to produce an absolutely accurate evaluation of the increase of the capital assets, but to identify the underlying structural limitations. For this reason, it is sufficient to group the entrepreneurs into rough ranges of high, middle or low income. Nevertheless, the reliability of the provided information needs to be ensured as far as possible. Therefore, entrepreneurs of the same profession have been visited within different target regions, which allows one to compare the results inter-regionally. In the case of dairy farming additional entrepreneurs were included through a group interview session in order to improve the sample size, which guarantees a higher credibility. At the end of the survey the findings were also presented to the leadership team of Drishtee to receive feedback regarding its authenticity. Finally, it will be proved that the provided information matches with relevant secondary source information regarding the income base within various professions.

The sectoral overview shows that, except for the dairy industry, the profession is closely connected with the BPL-status. Only one out of five ladies' shop-keepers was from an APL-background, while the stitching lady as well as the charkha workers and the handloom workers were all from a BPL-background. This is reflected in the monthly income, which was very low for the ladies' shop-keepers as well as the one textile worker, and between a low and middle level for the charkha workers, the handloom workers and the beautician. Most diverse was the group of dairy entrepreneurs, which ranges from BPL to APL-plus (although the APL-plus entrepreneurs are not considered in Figure 11). However, the high number of the APL-people among them was distorted by the ten entrepreneurs from the group session in Khagra, who all came from an APL-background (even though in that dairy class there were nine from BPL and twelve from APL).

Figure 11: Average monthly income, sector-by-sector for youth and adults (in Indian Rupees)[66]

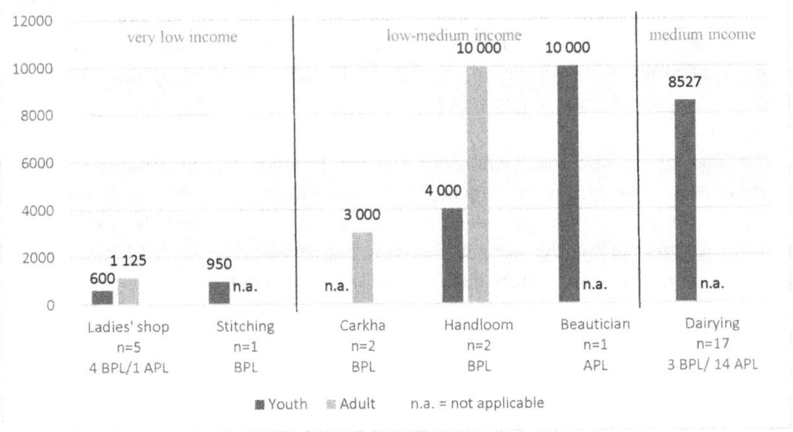

	very low income		low-medium income	medium income	
12000			10 000	10 000	
10000					8527
8000					
6000			4 000		
4000		3 000			
2000	1 125 / 600	950 / n.a.	n.a.	n.a.	n.a.
0					
	Ladies' shop / n=5 / 4 BPL/1 APL	Stitching / n=1 / BPL	Carkha / n=2 / BPL	Handloom / n=2 / BPL	Beautician / n=1 / APL	Dairying / n=17 / 3 BPL/ 14 APL

■ Youth ▨ Adult   n.a. = not applicable

Source: Author's own draft.

With regard to the provided income information, two features must be noted. Firstly, it has no claim to be representative for the respective sector, as the sample number is too small. And secondly, variations within all of these groups are possible, as the average income is dependent on many other factors, such as the time since the business was established, the village infrastructure, access to the market place and access to physical or natural capital assets, e.g. access to a fish pond, a charkha machine, a handloom or a stitching machine. Therefore, each case must be considered individually within the respective professional sector.

Now the different sectors will be examined in more detail. At first, only those whose increase of income cannot be compared before/after the training are concerned; this applies to the Rangoli shop-keepers, the charkha workers, the textile worker (who was still in training by the time of interview), and the handloom workers, as they conducted the training many years ago, and the respective training provider had not been visited.

The very low income of the five ladies' or Rangoli shop-keepers can partly be explained by the fact that the Rangoli-project was still in its pioneering stage. Two of the entrepreneurs opened their shop less than a year ago, and the other three opened theirs two to three years ago. In addition, the location in remote villages was certainly aggravating. Some stated that in a normal working day they have just one or two customers. However, this number can substantially increase during festive seasons, an aspect which was also noted by the stitching lady and the beautician.

The stitching entrepreneur (SE1) became self-employed in a tailoring shop for ladies around 15 years ago, in 1999. In contrast to most of the other entrepreneurs, she was running her business alone, which means that she had no family support and no employees. She calcula-

---

[66] 8000 Indian Rupees, which is approximately equal to 100 Euros. The exchange rate between the time of visit and now has changed only very slightly.

ted her monthly income to be 500-700 Rupees in the off-seasons, and 2 000 Rupees during marriage seasons.[67]

If a differentiation of these entrepreneurs' level of poverty were carried out, these charkha workers would have counted among the poorest. This is especially visible in their very low levels of education. One lady did not receive any school education at all, and the other went for five years to school. These two were both house-wives before and occasionally worked as labourers. They started this self-employed activity one and a half and two years ago respectively, and had been recruited for it by Drishtee, who also provided the training. According to personal information given by the Drishtee representatives, two women of the Murli village had been selected by Drishtee to be equipped with automatic running machines, due to their good performance. One machine ran from electricity, the other from solar energy (and was used by CE2). This was a great advantage in comparison to the hand-working machines (which CE1 was using). CE1 criticised that she felt tired from the manual machine and explicitly expressed her desire to become equipped with an electronic device too.

The two male handloom workers both came both from a traditional handlooming family. The adult handloom worker (HE1) registered that his monthly income was 10 000 Rupees, while the younger worker (HE2), who belonged to a different handloom family, estimated it to be much lower.[68] These two also had very low levels of education: the older one received five years of schooling, while the younger one never attended school.

6.1.2.2 The increase in business since the training

A comparison regarding the increase of business before and after the training is only applicable for those entrepreneurs who already had a running business at the time of training. This applied to most of the dairy entrepreneurs and the lady who was trained in beautification. In this section, a detailed look is first taken at the increase in income of the young dairy entrepreneurs as well as the young beautician entrepreneur (Figure 12). For the dairy entrepreneurs, it is reasonable to measure the increase in business not only by examining the increase of income, but also by looking at the number of cows or buffalo before and after the training (Figure 13).

---

[67] The figure of 950 Rupees given in Figure 12 is a rough estimate, assuming the off-season to be nine months long, and the marriage season to take up the remaining three months in a year (4:1 ratio). The same ratio was assumed for the beautician and stitching ladies.

[68] In case of the young handloom worker (HE2) the income has been calculated on the basis of his information that he earned 45-50 Rupees (approximately 60 cents) per meter, and makes 6.5 meters a day.

Figure 12: Increase in income following training for young entrepreneurs (in percentage and Indian Rupees)[69]

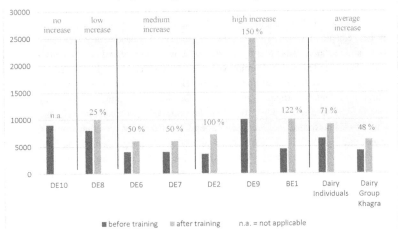

Source: Author's own draft. For the legend regarding the entrepreneurs, see Figure 10 (6.1.1.1).

Among the dairy entrepreneurs were six individuals (excepting DE5, because he was still in training at the time of interview), plus the ten additional entrepreneurs from the Khagra-group session. In accordance with the category system the following groups have been formed: "no increase", "low increase", "medium increase" and "high increase". It is notable that, except for the one with no increase (DE10), all entrepreneurs claimed that their income increased following the training. The ten additional entrepreneurs from Khagra all noted an increase in income, even though four of them did not get additional cows. This indicates that the dairy training course was an effective tool and poses the question of what specifically contributed to the increase of income. In order to find an explanation to this question, it is reasonable to consider this category on a case to case basis under the following aspects: first by asking when the training was taken, and second to inspect whether they purchased new cows or buffalos since the training was completed.

The training was completed by DE6 and DE7 from the Khagra-group in Bhagalpur only ten days before the time of interview, and by DE8, DE9 and DE10 from Munger five months earlier. DE2 from the Madhubani district had already completed it one year earlier, and BE1 at least three years ago. This could be an explanation for the high increase of DE2 and BE1. BE1 stated that the business had already become started ten years earlier.

For the dairy entrepreneurs, the increase in income since the training (Figure 12) is now considered in connection with the average number of cattle before and after the training (Figure 13). The ten surveyed dairy entrepreneurs (DE1 – DE10), four of which came from the Madhubani district, three from the Bhagalpur district and three from the Munger district, have all given exact information regarding their current stock of cows or buffalo, as well as their old

---

[69] 10 000 Rupees equal approximately 125 Euros.

stock before the dairy training started. This data allows us to evaluate the increase before and after the training. However, the information regarding the daily milk yield and the monthly income was only partly given. Most entrepreneurs provided information on either one or the other, but not both. Due to the small sample size, the number of cattle before and after the training is presented for each entrepreneur individually (Figure 13). The sample number was then increased through the two group sessions, which were conducted in the villages of Khagra (ten additional entrepreneurs) or Borwah (eight entrepreneurs, of which three came from BPL: five males and three females). Hence, the average number of cows of the dairy entrepreneurs from these groups are compared with the average number of the ten individual entrepreneurs. In the case of the entrepreneur with "no increase" (DE10), the current income was not applicable, as her family did not receive any income since the completion of the training, due to the pregnancy of the cow, but they would calculate the income anew, once they consume milk again.

**Figure 13: Number of cows or buffalo of the dairy entrepreneurs
(before and after the training)**

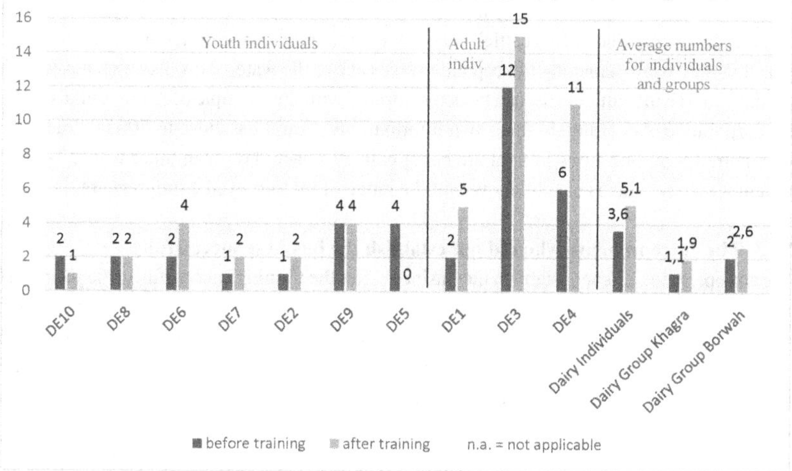

Source: Author's own draft. For the legend regarding the entrepreneurs, see Figure 11 (6.1.1.1).

The "low increase" of DE8 was obviously because he did not increase the number of cows. The two female entrepreneurs with a "medium increase" (DE6 and DE7) reflect, as part of the Khagra-group, the overall trend of that group, which had an average increase of nearly 50%. Even though they completed the training only ten days earlier, both of them said that they doubled the number of cows. Among the entrepreneurs with a "high increase", DE2 and DE9 deserve special attention: DE2 originally started the dairy business with only one buffalo, before he increased the number of buffalos to eight, but shortly after sold six of them again, because they were not giving much milk. So, right before the training he had only two, a statistic which had not changed since. Therefore, the calculation of income before and after the training is based on two buffalo. He explained that his income increased by 100% because the milk yield per buffalo had doubled. Before the training, one buffalo only gave two litres of milk, whereas now it gave four litres. Meanwhile, the case of DE9 (from BPL) is somewhat anomalous, as this entrepreneur was the leader of a farmers' club, who stated that his income grew after the training from 10 000 Rupees to 25 000 Rupees[70] due to his social engagement, even though the number of his cattle remained the same (this will be discussed under 6.1.3: Socio-cultural empowerment).

Taking into consideration the adult entrepreneurs above the age of 35, significant differences appear between those entrepreneurs who came from an APL-plus (DE3 and DE4) and those from a BPL/APL-background. It seems not to matter whether they come from a lower APL or BPL background, as both were engaging in dairy farming only on a small scale, and both represented entrepreneurs with a medium and high increase of income. In contrast, the two APL-plus candidates engaged in dairy farming on a larger scale and had more favourable conditions to extend this activity, as will be shown when taken in conjunction with these entrepreneurs' access to financial support.

The low average number of cattle, and consequently the low milk yields, were also confirmed by the Drishtee training manager in Madhubani, who said that "90-95% of the candidates have only one or two cattle". According to him they were doing only "so-so", which means that they were managing to keep themselves above the water. This also matches with the secondary source information, as can be demonstrated with the example of Amul, India's largest dairy corporative. According to their own information[71], Amul has close to 700 000 members, from whom 2.5 million litres in total are being collected each day, or in other words 3-4 litres per member a day. This is absolutely within the range of the surveyed entrepreneurs.

6.1.2.3 The entrepreneurs who did not establish the business successfully

Three entrepreneurs did not establish the business after the training successfully: the young lady who was trained in inland fisheries, but who did not get a loan to purchase a fish pond, and the two women trained by the Fateh Help Society in the Bhagalpur district, who in the meanwhile were pursuing different activities. These examples will now be looked at in contrast to the other entrepreneurs who had an active business, in order to understand the limiting factors that led eventually to the business failure. This analysis will be carried out through triangulation of the data, taken together with the insights from the three entrepreneurs on the one hand, and insights from the respective training representatives on the other hand.

Insights from the entrepreneurs
In the case of the fishery lady, she took the entrepreneurship training three years previously, but did not gain access to a fish pond, which she explained as being due to the scarce availability of fish ponds, which is confronted by a high demand. Therefore, she could not establish the business. As mentioned earlier, she was denied a loan due to the improper behaviour of the bank. It must also be noted that she was a mother of a two-year-old child at the time of interview,

---

[70] Approximately from 125 to 320 Euros.
[71] Published on the internet-page of Amul, http://www.amuldairy.com/index.php/the-organization/an-overview (Dec 2, 2017).

which may also have kept her from being focussed on the business.

In the case of the Fateh Help Society, a production unit for the making of different types of bags from jute was visited in a village of the Bhagalpur district. For this activity only one training batch was offered two and a half years ago, to prepare women to become self-employed. Seven women from the same village participated in it. However, this unit was not operative, but idle, for the last two years. During the visit the unit leader, who was also the trainer, as well as the two ladies who participated in the training batch as students, were interviewed. This production unit of the Fateh Help Society depended on this organization to provide them with materials in order to make bags, but for some unknown reason they stopped the provision. Hence, the two ladies, SE2 and SE3, were making these jute bags only for two months, until they had to stop. In explaining the circumstances SE3 expressed her dissatisfaction regarding the payment: "The goods that were required to make the jute bags were not sent by the Fateh Society, that is the main problem. If the Fateh Society sends goods required for preparing a bag, then I could work in a broad way. But Fateh Society does not give the goods at the right price to the required income, it is not in the range to get any income to improve it all. The labour cost is very low that I give. I get only 10-15 Rupees[72] per bag." With regard to getting a loan, SE2 explained that she did not try to get a loan as she thought she would not be able to return the money. She also mentioned that – apart from making jute bags – she was also involved in stitching. Thus, it is unclear whether she tried to apply for a loan for her personal involvement in stitching or for her engagement in the jute bag-making as part of the Fateh Help Society. In any case, both ladies were not able to find a way out of these difficulties, so the production unit remained inactive.

Insights from the training representatives

At this point, the perception of the unit leader as a representative of the Fateh Help Society is next taken into account. The unit leader blamed the bank for not being able to continue this jute-bag business. She explained: "Only seven girls, the loan amount for them got just ignored, because they would be unable to return it back. That is the main point, that is why I cannot take a loan and open any self-employment business. If I want to have my own business then I need to have at least 35-40 000 Rupees[73], but then I need to have the materials and I am afraid of taking the risk." The unit leader further explained that there was a SHG in the village for self-employed women. They collected 50 Rupees[74] a month from each person. Other women whom she trained were still part of that group and received loans, which meant that others from the same training batch must have been successful in opening other types of self-employed activities, if they were part of the group and acquired money.

In conclusion, what led to the business failure in these three cases is a lack of access to physical as well as financial capital, e.g. the materials for making bags or the fish ponds, for which the entrepreneurs relied on a loan which was not granted to them. With regard to their economic empowerment, they were all restricted by financial and other organizational constraints, but

---

[72] Approximately 12-20 cents.
[73] Approximately 445-510 Euros.
[74] Approximately 60 cents.

chose to accept these limiting structural conditions instead of developing a personal initiative to confront these constraints.[75] SE2 and SE3 relied severely on the training provider and adopted its passive behaviour concerning the acquisition of financial resources, while FE1 relied on the behaviour of the bank.

6.1.2.4 Access to financial support

The survey revealed that access to a loan from a bank was one of the most distinctive examples in which striking differences between rich and poor became visible. Again, insights on this issue are gained through the means of a triangulation of the data, taken together with insights from the entrepreneurs and training representatives.

Insights from the entrepreneurs

Figure 14 shows the extent to which the surveyed entrepreneurs gained access to financial support. In this light, the adult entrepreneurs above the age of 35 are compared with the young ones. The relevant subcategories include "gaining access to a loan from a bank", either as an individual person or through a farmers' club, "borrowing money from family members or friends", or "receiving a start-up loan from the training provider". This category only applies to the Rangoli shop-keepers, who equally received a loan of 20 000 Rupees[76] from Drishtee as part of the start-up package. It is also shown who wanted to apply for a loan or who deliberately decided not to apply for one, and who tried to get a loan but did not receive one. There is, in addition, one case which did not occur in the survey, and therefore seems to be missing, that is borrowing money through a middle man. This is a common practice in rural parts of India and is often a transaction arranged at only disproportionate interest rates (see Sarmistha 2015).

On the one hand, there were three persons from APL-plus, who all received a loan directly from the bank. The two farmers trained by RSETI in dairy farming (DE3 and DE4) were promised by the creditors to receive a loan if they took the dairy training, which was an incentive for them to participate in it. They both used the money granted to them to purchase more cows or buffalo. VE1, the APL-plus entrepreneur from the Madhubani district who was trained by Drishtee, received a loan trough his KCC (Kisan Credit Card), a farmers' credit card which is a product designed by NABARD to enable farmers to gain access to credits.

---

[75] The importance of the development of personal initiative will be further considered under 6.2.
[76] Approximately 250 Euros.

Figure 14: Access to financial support

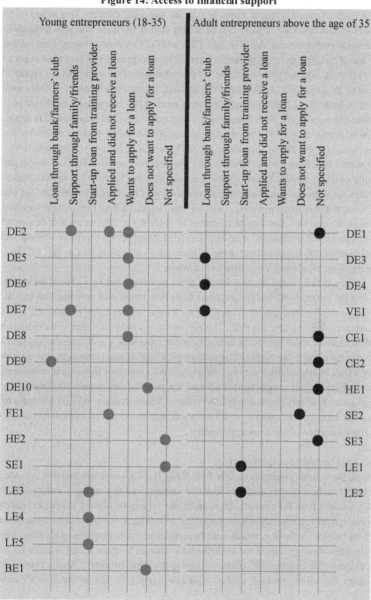

Source: Author's own draft. For the legend regarding the entrepreneurs, see Figure 10 (6.1.1.1).

115

On the other hand, among the remaining entrepreneurs, only DE9 already gained access to credit through a farmers' club, of which he was the managing director. This means of getting involved in a farmers' club or SHG will be further considered under the section focussing on socio-cultural empowerment below (6.1.3). DE2 and DE7 borrowed money from relatives or village members. Moreover, a few people from a BPL-background had also tried to get a loan. FE1 said that she was not able to get one as "the behaviour of the bank was not proper". DE2 also tried to get a loan, but simply stated that he faced "some complications". Others were still planning to apply for a loan. Among them were most of the young dairy entrepreneurs. It is noticeable that none of the elder entrepreneurs explicitly stated that they wanted to apply for a loan. Other people from BPL did not even try to apply for a loan, such as SE2, who doubted that she would be able to pay the money off. DE10 and BE1, both of whom came from APL, did not want a loan from a bank. Whereas BE1 was focussed on expanding her beauty parlour, the family of DE10 decided to not expand their dairy farming business any further. There were two professional groups that were not assigned to one of the categories, namely the charkha workers and the handloom workers. They may have wanted a loan from a bank, but they did not specifically mention that they also wanted to apply for it. This seemed to be a hurdle to them. It appears that the Rangoli shop-keepers relied altogether on the initial start-up package that included 20 000 Rupees. Since some of them were still paying this money off, they also did not explicate their interest in a loan from a bank.

In conclusion, it can be said that the grant of a loan is a very crucial instrument for expanding the business. All the entrepreneurs from BPL relied on a very minimal capital base and were generally very cautious in applying for a loan. which might be a hint at the fact that they were not willing to take any risks. The question now appears whether, if the entrepreneurs from BPL received more assistance in how to apply for a loan, e.g. by the training providers, they would have had more courage and applied for a loan. If this is the case, it is certainly an issue of a lack of empowerment, if they simply did not know how to go about it and were not informed about certain programs that address the poor to gain access to financial capital. Another question at this juncture would be whether they actually would be granted a loan if they applied for it. Further insights on this issue will next be introduced through the perspectives of the representatives of the training providers.

Insights from the training representatives

In general, the training managers (except from RSETI) did not mention that they provided any active help as to how to gain access to financial support. Both the manager and teachers stated that they would advise their trainees to go and apply for a loan. A dairy teacher in Madhubani mentioned not only that he advised the trainees to go to a bank, but also that he issued a request to the bank to approve the loan, which would mean that he was willing to provide active support. Yet, only a few were specific about available loan schemes or about where to apply for a loan. The manager of the Drishtee/ITC centre referred to the Prime Minister's Mudra loan scheme.[77] Some dairy as well as stitching teachers instead referred generally to the PMKVY-scheme,

---

[77] MUDRA – Micro Units Development and Refinance Agency, a loan scheme launched by the Prime Minister of India in April 2015.

through which any student who successfully completes a PMKVY-training course will get a loan of 8 000 Rupees[78] from the government. However, with respect to the question of finding assistance to get access to financial support, which goes beyond the duty of the trainer, any kind of handholding with the graduates is not encouraged. For example, the stitching teacher of Drishtee in Madhubani said that the government would give the students a loan for purchasing stitching machines. But she would not support them in applying for a loan. As for the specific numbers of how many students have been given a loan and from which type of institution, only the manager of the Drishtee training centre in Madhubani was able to provide specific information and mentioned that NABART, as well as the State Bank of India, selected 62 of their trainees in dairy farming, granted them a loan and linked them with a milk production company named "Sudar".

When asked about the bank's requirements to obtain a loan, most of the training representatives had to admit that they were not aware of them. This was the case for the training mobilizer of Drishtee or a dairy teacher in Bhagalpur, who said that while 90% of the students graduate, only 10% of those who graduate will get a loan from a bank. The manager of the Drishtee training centre in Munger, moreover, stated that "banks are often not ready to pay. They require applicants to have their own property" and advised his trainees to go for it if they are "eligible" for it. According to DT5 from Munger, it was due to the trainees themselves that they were not getting loans, and he explained "20% of all of the students will take a loan from the bank, but ... they have to pay lots of interest to the bank. That is the reason they all are not taking a loan. They try to get self-employment from other sources of income."

As for the entrepreneurs who were trained through RSETI, every RSETI was in touch with a bank, which was the Central Bank of India in the case of the RSETI in Madhubani. This RSETI manager referred to the fact that every bank follows a procedure in which they prove if a candidate is eligibility for a loan or not. He then added that they were making arrangements to meet the students who successfully completed the training and to help them to build their relationship with the bank. Yet the survey showed that not everyone who had completed a RSETI training received a credit, such as the lady trained in inland fisheries, who noted that the behaviour of the bank was improper. Here, the question remains open whether RSETI assisted that young entrepreneur to get a fish pond and apply for a loan or not.

With an eye on the curricula, any type of entrepreneurship training includes marketing sections (compare 2.2.3.7 and appendices I-V). However, interviews with both the entrepreneurs and the trainers have shown that practical advice concerning where to avail support, and how to apply for a loan, was not given in the class. However, the picture that the representatives of the training providers painted still matches with the reality that the surveyed entrepreneurs have reflected. The assumption that entrepreneurs from poor backgrounds are very disadvantaged compared to wealthier candidates was on the whole upheld, especially if the entrepreneurs concerned did not own any property. This indeed seems to be a barrier for many to apply for a loan. But it also revealed that, after completion of the training course, the entrepreneurs were, by and large, left to themselves in taking the initiative to apply for a loan. It seems, then, that support and assistance is needed not only in order to apply for a loan, but also in getting engaged

---

[78] Approximately hundred Euros.

in a farmers' club or SHG, which is without doubt a door-opener to financial support. This issue will be elaborated in connection with the consideration of their socio-cultural empowerment under 6.1.3 below.

6.1.2.5 Applying training inputs

Next will be considered a key category for the analysis, namely the application of training inputs. In agreement with the first specified research question and its associated hypothesis, the goal of this section is to understand how the graduates managed to utilize specific training inputs of the entrepreneurship training course, in order to gain or increase access to various capital assets. The last part, relating to gaining or increasing access to capital assets, was previously considered in the evaluation of the entrepreneurs' access to financial and physical capital, in connection with their empowerment through the entrepreneurship training. Now the main attention is placed on the first part, the utilization of specific training inputs. In the theoretical framework, the process of empowerment was discussed in connection with Giddens' theory of structuration (2.1.2). This way, emphasis was placed on the *learning* which takes place through the entrepreneurship training, and its effect on the practical and discursive consciousness. It can be assumed that, for most of the entrepreneurs, the exercise of the self-employed activity had become a routinized practice. Yet again, after a period of time, reflexive knowledge had been acquired through the training course, which increased the entrepreneurs' discursive consciousness. Now the underlying assumption was that it would be possible to apply this acquired reflexive knowledge to the self-employed activity in such a way that the agency would *thicken* over time. Or in other words, the ability to act would increase, and with that the entrepreneur's capability to cope with structural limitations would also go up.

In order to identify these effects through the application of the training inputs to the self-employed activity, Figure 15 (denoting the young adults) and Figure 16 (denoting the adults) itemize the acquired discursive knowledge (as illustrated by the yellow knowledge box) into two aspects. Firstly, a differentiation is made between the theoretical knowledge and the practical application. Thereby, the entrepreneurs' self-perception is captured. Did all their knowledge remain in the knowledge box, or did they apply it to the self-employed activity? Secondly, a differentiation is made whether the things that have been learnt have been verbalized in a more general or a more specific manner. The more specific the entrepreneurs talk about what they have learnt and applied and what kind of results it had, the greater it would indicate that the training course had a direct impact on the entrepreneur's agency or ability to act. Whereas some entrepreneurs only verbalized what they have learnt, others also verbalized how the situation was before, what they were more specifically able to apply, and how it became visible in practice.

In a next stage, these perceptions are linked with the previous findings from the category-based analysis regarding the increase of business, which serves as an indicator whether the business success can be explained though the application of training inputs. For the young entrepreneurs, a link is made to the "increase of finances since training". Since this category has not been evaluated for the older entrepreneurs (above the age of 35), it seemed reasonable for them to consider the utilization of training inputs in connection with their social status, as is determined in terms of not only their APL/BPL-status, but also their educational level. Furthermore, the older entrepreneurs did not explicitly verbalize how the situation was before the training, and so the middle box representing this is missing in Figure 16. For both the young and older entrepreneurs, it must be noted that not everyone provided information regarding this category, such as the charkha workers, the handloom workers and the handbag-makers. For the charkha workers, the effect of the entrepreneurship training is obvious, as the training course was a prerequisite element to getting their self-employed activity started.

After the consideration of the application of training inputs, it seems appropriate to draw attention to the entrepreneurs' evaluation of the entrepreneurship training. This will be achieved in connection with the previous discussion regarding the application of training inputs.

Figure 15: Application of training inputs in connection with the increase of income for the young entrepreneurs (18-35 years)

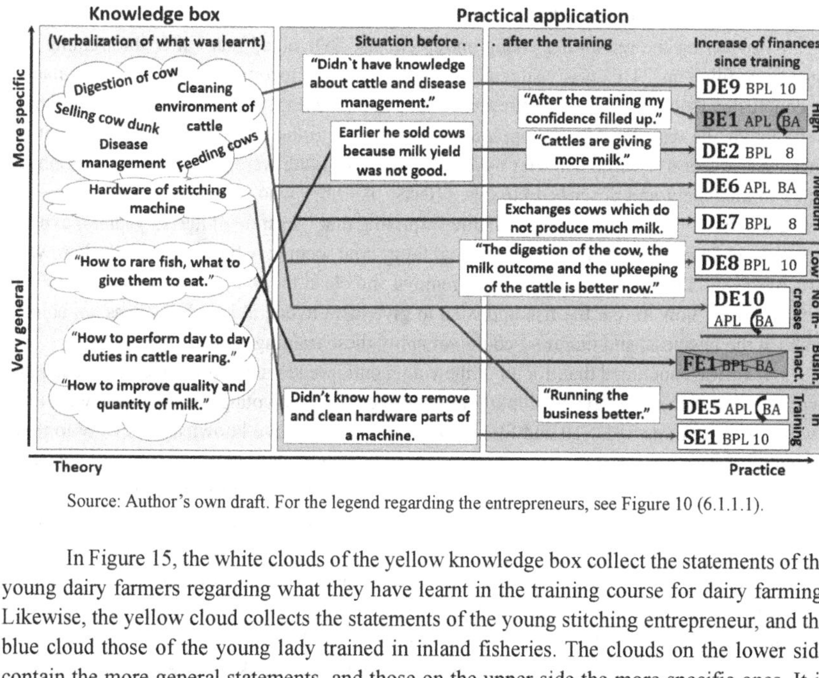

Source: Author's own draft. For the legend regarding the entrepreneurs, see Figure 10 (6.1.1.1).

In Figure 15, the white clouds of the yellow knowledge box collect the statements of the young dairy farmers regarding what they have learnt in the training course for dairy farming. Likewise, the yellow cloud collects the statements of the young stitching entrepreneur, and the blue cloud those of the young lady trained in inland fisheries. The clouds on the lower side contain the more general statements, and those on the upper side the more specific ones. It is noticeable that all specific statements regarding the things which have been learnt through the entrepreneurship training – except for "selling the cow dung" – relate to technical knowledge, e.g. the digestion of the cow or the hardware parts of the stitching machine. None of them referred to any of the entrepreneurial aspects included in the respective curricula, e.g. market knowledge or building customer relations. However, for all entrepreneurs except for DE9 and DE10, it seems plausible to explain their increase in income – apart from the fact, that some of them increased their number of cattle – by their application of training inputs. DE2 doubled his milk outcome per cow through the training, even though he was not very specific, and simply stated that "the training enlarged my knowledge about how to perform day to day duties in cattle rearing". According to him, this caused his cattle to give more milk. DE8 did not increase his number of cattle, but still improved the milk yield. He was very specific in terms of what he learnt and what he was able to apply, stating that he learnt that "the everyday intake of salt should not increase more than 50g". Furthermore, he explained that after the training "the digestion of the cattle has increased, so the productivity of milk has increased, and the upkeeping of the cattle is better now". This means that the entrepreneurs' 25% increase of income is due to a better understanding and application of what food to give to the cow, which resulted in the increase of milk.

In contrast, for DE9 the high increase of income cannot be explained by the application of training inputs, but by his involvement as a manager of a farmers' club. DE10 is an example where all the acquired knowledge had to stay in the knowledge box for the moment, as they were not able to get milk from the cow due to her pregnancy.

In viewing the remaining young entrepreneurs, BE1 noted that "after the training my confidence filled up". This was visible in the high increase of income, which allows the plausible conclusion that she developed entrepreneurial competencies, which filled up her confidence, even though she was the only entrepreneur who referred to the training effect without noting what she had learnt. Even those who were currently in training were able to apply the training inputs immediately and described positive effects, like DE5 who said that "I am running the business better". For SE1 it appeared a little surprising that – in spite of her 15 years of experience – she claimed that she did not know about things that seem very basic, such as the hardware parts of the stitching machine, and how to remove and clean them. FE1 learnt all the practical parts regarding how to rear the fish and what to give them to eat, although she was not able to establish the business, and therefore could not apply these training inputs.

It can be concluded that, for all of the young entrepreneurs, the reflexive knowledge did increase through the entrepreneurship training. Furthermore, the young entrepreneurs were able – with the exceptions of DE10 and FE1 – to apply their reflexive knowledge directly to their self-employed activity. For most of them, this was visible by an increase of income, i.e. DE8 had a low increase, DE6 and DE7 a medium increase, and DE2 and BE1 a high increase of income. Only in the case of DE9 was the high increase of income directly linked with his involvement in a farmers' club. This supports the first hypothesis that the young entrepreneurs' agency *thickened* as a direct result of the entrepreneurship training. However, the increase was only by a very limited degree. It did not significantly change their economic conditions, e.g. by bringing them out of poverty. In this way, it is notable that only technical knowledge was named in the entrepreneurs' statement, but nothing related to marketing skills, even though these aspects were part of the curricula. In addition, the entrepreneurs' statement did not refer to the further structural set-up, e.g. cooperatives, farmers' clubs or SHGs. Only DE9 seemed to have a broader understanding of how to avail the support structure.

Figure 16: Application of training inputs for adult entrepreneurs (above the age of 35)

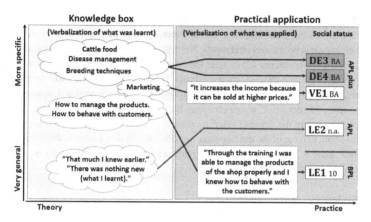

Source: Author's own draft. For the legend regarding the entrepreneurs, see Figure 10 (6.1.1.1).

Figure 16 views the application of training inputs for the adult entrepreneurs in connection with their social status. The three APL-plus candidates – all of whom were very wealthy farmers with large estates and diversified cultivations – had in common the fact that they acquired the entrepreneurship training at various training locations from different institutions (see for further details the considerations regarding the development of personal initiative under 6.2), which means that they had a very high level of expertise. Yet, they regarded the training as a good experience and named specific things they have learnt. For DE3 and DE4, participation in the training was directly connected with gaining access to a loan, due to the close relationship between the RSETI and the credit institution. It is notable that only VE1 explicitly referred to entrepreneurial aspects, i.e. marketing in connection with vermicomposting, while DE3 and DE4 highlighted only technical knowledge. For the Rangoli shop-keepers it is obvious that, with respect to applying training inputs, any generated income was a direct result of the 21-day-long entrepreneurship training, as participation in it was a prerequisite to open a shop. Nevertheless, the entrepreneurs had very contrasting views as to what extent they benefited from the training. LE1 said that "through the training I was able to manage the products of the properly and I knew how to behave with the customers", whereas LE2 said regarding the application of training inputs, "that much I knew earlier, there was nothing new". It is also highly notable that LE1 added a beauty parlour to her business, which was not part of the training curriculum for opening a Rangoli shop.[79] Even though she did not make a greater income than the other Rangoli shop-keepers, she stated that she completely reinvested all the income into this advancement and did not spend any of it for her personal use.

In conclusion, the adult entrepreneurs – if one leaves the subjective perception of LE2 aside – increased their reflexive knowledge through the entrepreneurship training, and were

---

[79] Although she took a training in beautification with a different training provider.

able to apply it to their self-employed activity, irrespective of their BPL/APL-status. Yet, the BPL candidates remained on a very low level, whereas the APL-plus candidates from RSETI benefitted from the training disproportionately, as for them the participation in the training was directly linked with the grant of a loan, which explained why they were able to significantly expand their business operations afterwards. Nevertheless, the understanding of why both the young and adult entrepreneurs coped with their structural limitations very differently seems significant with respect to the first two hypotheses. Hence, further insights will next be drawn with regard to the way that the entrepreneurs manage to cope with structural conditions. This will be completed under 6.1.3, which considers their socio-cultural empowerment, and under 6.3, which focusses on the entrepreneurs' awareness of enabling or limiting structural conditions. In addition, the role of the development of personal initiative has to be discovered, which will be done under 6.2. Only then can the hypotheses for SQ1 and SQ2 finally be verified, a stage that will be reached under 6.3.

6.1.2.6 Evaluation of the training course

In order to obtain a subjective view on the entrepreneurs' opinions regarding the entrepreneurship training, they were asked to grade the training course between A (very good) and D (insufficient), as is illustrated in Figure 17. These insights are intentionally placed after the analysis of the application of training inputs, in order to find out whether the positive effects of the training course are reflected in the grading of the training course or not. This thought-provoking question also inspired some of the entrepreneurs to ponder the strengths and weaknesses of the training course, and it thus provides insights into their awareness of any shortcomings. Apart from a few exceptions, all entrepreneurs are included who had already completed the training at the time of interview, as well as those for whom a representative of the training provider has been interviewed.

Figure 17: Grading of the training course between A (Very good) and D (Insufficient)

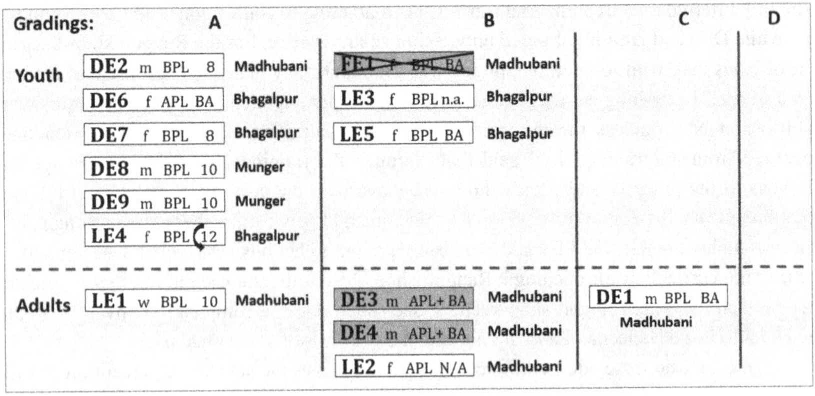

Source: Author's own draft. For the legend regarding the entrepreneurs, see Figure 10 (6.1.1.1).

122

Out of 14 respondents, seven graded the training course with an A grade, six with a B grade, only one person with a C grade, and none with a D grade. This picture shows that, irrespective of their educational or social status, they were individually able to make a judgement without knowing what others from the same training batch have said. Nevertheless, they were very restrained about mentioning criticisms regarding the training course, which means that they either were not aware of it or they were holding it back. Only DE1, who graded the training course with a C grade, gave very precise and valuable criticisms by saying "theoretically everything was done. People were taught. People were trained. But the practical part was not done... like supporting the farmer". In this criticism he went on to mention some aspects which should have been covered by the training, such as mental support, economic support, and marketing support. Moreover, it is worth noting that DE2 – who graded the course with an A – and DE1 – who graded it with a C – were in the same training batch one year previously. Even LE2, who said that she did not learn anything new in the training, graded it with at least a B. It appears that all the entrepreneurs, who mentioned a positive before/after-training-effect in Figures 15-16, have graded it with an A, i.e. DE2, DE5, DE7, DE8, DE9 and LE1. This fact proves the close link between the positive effect of the entrepreneurship training on the self-employed activity on the one hand, and the generally positive grading of the training course on the other hand.

Overall, regarding the economic empowerment, it can be said that the entrepreneurs widely benefited from the entrepreneurship training and were able to extend their businesses to a limited degree. This was the case even for those who had already established their business and had a lot of previous experience, like many of the dairy entrepreneurs or the textile worker. On the one hand, practical training inputs were immediately applied, such as the right nutrition and disease management of a cow, technical know-how regarding the use of working equipment, e.g. a stitching or a charkha machine. On the other hand, management decisions were taken as a result of the entrepreneurship training, e.g. the purchase of cows. And yet, most of the entrepreneurs with microenterprises behaved quite reactively, which means that they managed to apply the training inputs and, as a result, gained direct benefits, albeit to a limited extent, whereas whatever lies beyond the curriculum, e.g. engagement in social networks, applying for a loan, etc.[80] they reacted as being a lost cause and left this aside from themselves. Hence, gaining access to financial support remains a major constraint. This issue touches on the topic of building up social capital. This will be now looked at in more detail in consideration with the socio-cultural empowerment of the entrepreneurs, an aspect which is closely linked with gaining access to financial capital, and therefore with their economic empowerment.

6.1.3 Socio-cultural empowerment
In the context of the socio-cultural empowerment of young entrepreneurs, social networks play a vital role, specifically those connected with the entrepreneurial activity, since in the village

---

[80] Even though these aspects may have been mentioned in the curricula (see Appendices I-V), the trainers in general did not offer active support in these matters (compare 6.1.2.4 and the conclusion regarding handholding with the training provider below).

structure of rural India there are strong ties not only to one's own family and relatives, but also to the village community. These network relations are viewed in consideration with the impact of different socialization agents on the business activity. In compliance with the specified research questions 3, 4 and 5, the family, the training provider and the community-based organizations are considered as part of the structural set-up as primary, secondary and tertiary socialization agents. The following evaluative categories are now captured by means of a data-triangulation, with insights drawn from both the entrepreneurs and the training representatives.

6.1.3.1 Impact of the family as primary socialization agent on the self-employed activity

The family background of the young entrepreneurs (those aged 18-35) is now considered as part of the structural set-up.[81] As it was already outlined, the family can function as an agency thickener or thinner for the surveyed entrepreneurs (2.1.2). Therefore, a guiding question is to ask what impact the family as the primary socialization agent has on the entrepreneurial activity. Attention is drawn to the socio-economic set-up of the families of the entrepreneurs and the role that the self-employed activity plays within the household portfolio. Thereby, aspects of the family tradition and family support with regard to the business activity are considered.

Insights from the entrepreneurs and trainees

Figure 18 captures which of the young entrepreneurs received support from the family in the business activity. By support, this means in practice that the business was either considered as a family business, or that other family members helped out in the business. In addition, it shows any further activities or responsibilities of the entrepreneurs besides the business of training, such as further income-generating activities, if they were still pursuing education, or if they had responsibilities as a mother for nurturing children. Finally, it also shows further income-generating activities of the father/mother or the husband/wife. Thereby, the marital status as well as the BPL/APL-status have both been kept in mind.

In considering the household portfolios of the young entrepreneurs, three groups can be identified. Firstly, there were those families for whom the self-employed activity was the only activity and part of a long family tradition, e.g. in the case of the handloom families. Secondly, there were those families with a diversified household portfolio, but for whom the self-employed activity was also part of a long tradition, e.g. in the case of the dairy farmers. Finally, there were those entrepreneurs who became newly introduced to the self-employed activity through the training provider. In this last respect, this activity augmented the household portfolio as an additional income-generating activity. As for the first group, stark contrasts are visible between different traditional handloom families that – to a large degree – solely focussed and depended on this one household activity. In the case of the surveyed handloom entrepreneurs, their activity had such a high priority that school education seemed to have no importance in the eyes of the parents. Hence, other activities had not been viewed as a viable option for their children, an aspect which tremendously limits their agency and puts significant constraints on their ability to make their own choices. HE2 stands as an example of this as a

---

[81] At this point, only the young entrepreneurs are considered as the main target group, because for them the families impact is more obvious than for the adults, since they were still transitioning or having just accomplished the transition into adulthood.

son from a traditional handloom family. He said that he never attended school, but learnt the art of handloom from scratch, which consequently became his one and only source of livelihood. In contrast, a young lady of 18 years, who also came from a traditional handloom family from the same village as HE2, namely LE4, and who was interviewed as a Rangoli shop-keeper, attended the twelfth class in school, even though she came from a BPL-family and had six brothers and four sisters. She mentioned that her mother and sister were taking care of the shop when she was in school. These two cases confirm the high impact of the family on the self-employed activity, yet in each case the family has impacted the entrepreneur's agency in a different way. Whereas HE2's only option was to restrict his attentions to the family tradition in a low income-generating occupation within a declining sector (see 2.3.3.2), LE4 could thrive for other options due to her educational achievements. In light of this, it must be mentioned that handloom is a male-dominated occupation, which might have contributed to the parents' decision to send LE4 to school. In contrast to the handloom entrepreneurs, the family impact seemed to be very low in the case of SE1, an experienced stitching entrepreneur of 33 years with a daughter who was 17 years old, as she did not get any support from the family in her business activity.

For the second group, the dairy farmers, the business can be considered as a traditional family business, which means that it was already active before taking the training. However, with the exception of DE9, who was totally committed to dairying and managing a dairy farmers' club, this was an activity performed in combination with other activities. Here, DE2 stands as an example for a BPL-person with a diverse business portfolio, for which he seems to make management decisions very independently. He was already engaged in various previous low income-activities, namely in selling newspapers and running a small kiosk. At present, he was also running a general store, besides engaging in dairy farming. His decisive behaviour also became more visible in his decision-taking regarding the dairying business, as he increased the number of buffalo from one to eight, and, shortly after, decreased it again to two. He did not mention any family support, but it could be assumed that his family engaged in cattle-rearing as well, as he was most of the time in the shop. As for D10, the young lady from APL who was doing her Bachelor's degree in History, she also received support from the whole family, yet dairy farming was only a subordinate activity, as they were engaged in farming and her mother worked as a government teacher. The high support of the family generally was also confirmed by the interviews with the eight trainees, who were currently undertaking a training in dairy farming. All of them came from dairy farmers' families, which means that they already had cattle at home. But their families all practised dairy-farming on a very small scale in combination with other activities, either with farming or with non-agricultural activities, in most cases with handicrafts.

In the last group are only the three young Rangoli shop-keepers, LE3 to LE5, all of whom came from BPL-families, yet from different occupational backgrounds. LE3 came from a small village nearby the handloom village of Puraini. Her father worked as a labourer and her husband as a farmer. LE5 came from a charkha family, which was not part of their family tradition. All of them had the support of the family in running these shops, which enabled the younger entrepreneurs, LE4 and LE5, to pursue their school or university education. Yet the fact that they were engaged in such a low income-generating activity – in spite of their educational attainments – was certainly a sign of family constraints due to depriving economic conditions, from which they could hardly escape.

Figure 18: Household portfolio of the young entrepreneurs (18-35 years)

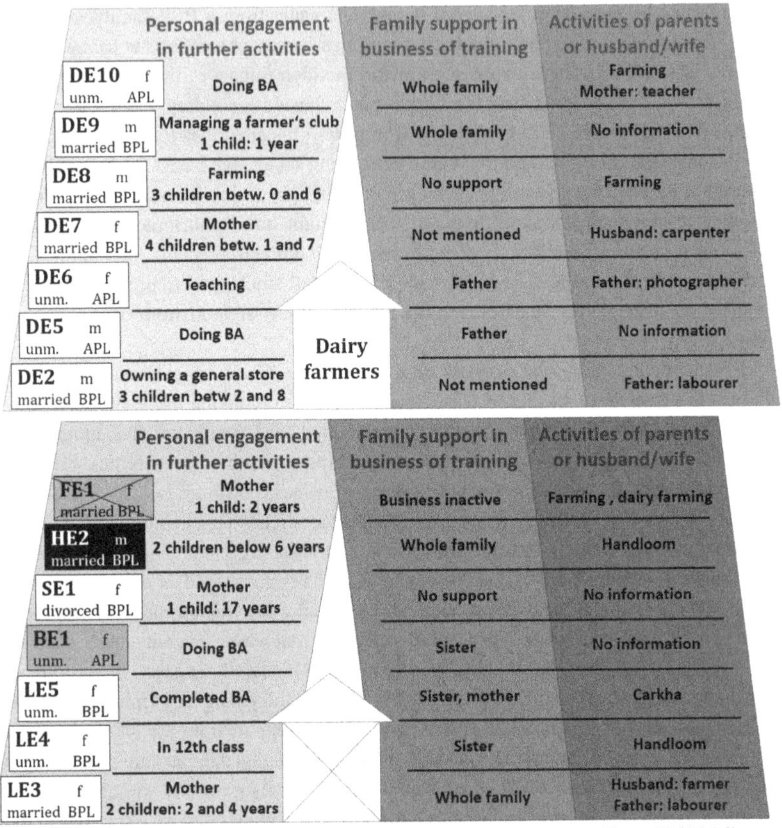

		Personal engagement in further activities	Family support in business of training	Activities of parents or husband/wife
DE10 f unm. APL		Doing BA	Whole family	Farming Mother: teacher
DE9 m married BPL	Managing a farmer's club 1 child: 1 year		Whole family	No information
DE8 m married BPL	Farming 3 children betw. 0 and 6		No support	Farming
DE7 f married BPL	Mother 4 children betw. 1 and 7		Not mentioned	Husband: carpenter
DE6 f unm. APL		Teaching	Father	Father: photographer
DE5 m unm. APL		Doing BA	Father	No information
DE2 m married BPL	Owning a general store 3 children betw 2 and 8		Not mentioned	Father: labourer

Dairy farmers

		Personal engagement in further activities	Family support in business of training	Activities of parents or husband/wife
FE1 f married BPL	Mother 1 child: 2 years		Business inactive	Farming, dairy farming
HE2 m married BPL	2 children below 6 years		Whole family	Handloom
SE1 f divorced BPL	Mother 1 child: 17 years		No support	No information
BE1 f unm. APL		Doing BA	Sister	No information
LE5 f unm. BPL		Completed BA	Sister, mother	Carkha
LE4 f unm. BPL		In 12th class	Sister	Handloom
LE3 f married BPL	Mother 2 children: 2 and 4 years		Whole family	Husband: farmer Father: labourer

Source: Author's own draft. For the legend regarding the entrepreneurs, see Figure 10 (6.1.1.1).

The active support of the family in the business was notable for all youth in transition – both those who were not married and those who had not completed their school or university education yet – irrespective of whether they came from an APL or BPL-background. Mostly individual family members were engaged in running the business, usually the father or mother. This raises the question of how much freedom these young entrepreneurs really have to change habits in the business activity in order to increase the income, as they were applying the training inputs. In the case of BE1, the sister had also taken the training in beautification. They were running the beauty parlour together, so they could be considered as the main entrepreneurs. She did not mention any further support from the family in the business, but exclaimed that her motivation for running this business was to support her family.

Insights from the training representatives

At this point the perspectives of the training representatives are considered in order to obtain a stable picture of the structural limitations related to personal and family constraints.

With respect firstly to the handloom workers in the Puraini village, where the two contrasting cases occurred – namely HE2, who never attended school, and LE4, a daughter from a handloom family from the same village who attended school up to the twelfth grade – the textile manager of Drishtee noted, during a home visit of HE2, that "many children are leaving the village. But (pointing to HE2, the son who was interviewed) he stayed and continued the family's business. He learned it already as a child". It must be emphasized that the three persons who have been interviewed from three different families of the Puraini village are only individual cases, and cannot be said to be representative of the whole village. But they each reflect a part of the local reality. On the one hand, the children are following the family tradition, in not having moved beyond a low educational level. On the other hand, they are leaving the village in search of other and more lucrative job opportunities.

The interview with the Drishtee mobilizer in the Bhagalpur district, who was responsible for the recruitment of the trainees, and who said that he has contact with 80 villages in various blocks of the Bhagalpur district, with a success-rate of 80%, revealed that the main reason for not participating in the training is a lack of education: "Due to a lack of education, they don't see any point of taking the training." This means that, for some individuals, personal and family constraints are a barrier to taking advantage of the entrepreneurship training in order to become engaged in a self-employed activity or to increase an existing activity, while for others, these constraints are keeping them from attaining a higher educational degree, as they are compelled to engage in the self-employed activity.

With regard to the specified research question 3, it can be concluded that the influence of the family on young entrepreneurs to become involved in self-employed activity is very high. This is supported by the fact that – with the exception of the stitching lady in Madhubani – the business was usually not performed alone but with the active support of the family. This means that for those who undertook the entrepreneurship training in an occupation that was already part of the family tradition, the initial impulse to become engaged in the self-employed activity was certainly given by the family and not by the training provider. If the occupation was new to the family, as in the Rangoli shop-keepers or the charkha workers, the initial impulse was given by the training provider, yet the young entrepreneur was still pushed by the family to become more involved in the activity as the son or daughter, a role which is often an obstacle for them to pursuing their school or university education at the same time. Nevertheless, the fact that a high share of the young entrepreneurs attained high educational degrees, even some females from a BPL-background, as e.g. LE4 and LE5, shows that some families are giving high priority not only to the self-employed activity, but also to the children's education, although not without incorporating this activity into a timetable that also includes domestic responsibilities.

6.1.3.2 Handholding with the training provider

After considering the impact of the family as primary socialization agent on the self-employed activity, the role of the training provider itself as the secondary socialization agent now needs to be emphasized. In viewing the training provider under the theoretical considerations laid out by Giddens' structuration, its function to provide enabling structures must be understood, which helps to extend the entrepreneur's agency. The role of the training provider as a motivator to participate in the entrepreneurship training has already been clearly demonstrated in the previous section (6.1.3.1). At this point, attention is drawn to the question whether the training provider also has an enabling role after the completion of the training course. Does the training course provide any handholding for the graduates? Not only that, how do they benefit from the assistance and the connections of the training provider (specified research question 4)? First, the perspectives of the entrepreneurs, and then the perspectives of the representatives of the training providers, will now be looked at.

Insights from the entrepreneurs

Figure 19 illustrates the handholding between the entrepreneurs and the training provider. The following differentials – from "very satisfying support/advice" to "support diminished / stopped" or "no support" – have been made as a result of the interviews with the entrepreneurs. The division "only stays in touch when necessary" means that the basic support to keep the business running is given by the training provider, e.g. the supply of a Rangoli shop with products, but no further assistance is provided. In all cases, the subjectivity of the entrepreneurs must be taken into account, which is why the perspectives of the training providers must be considered equally in order to allow one to draw a more holistic and coherent picture.

The RSETI entrepreneurs reveal two very contrasting views. On the one hand they were very satisfied, while on the other hand they were dissatisfied with the handholding with the training provider. One of the dairy entrepreneurs (DE3) from an APL-plus background seemed very satisfied when he explained that they come and visit and give advice. The same applies to the beautician lady (BE1) from APL, who said that still today they would come and give advice. On the other hand, the fishery entrepreneur from BPL, who did not get her business going due to her financial constraints, would have needed the support from RSETI in order to get settled, but she said that she did not receive any support.

The Drishtee entrepreneurs also paint a divergent picture. While DE8 said he "feels honoured to get information and help from Drishtee", DE1 wished to receive advice and support of Drishtee and explained: "Yes, I want advice from Drishtee so that Drishtee staff comes here, advice in cattle rearing and farming, in the affairs of economic activities." He answered the question whether he is getting all this with a simple "No". Yet, the question remains open whether he actually tried to receive this kind of advice, or if he simply waited for Drishtee to make the first step. This shows a similar set of circumstances to DE2, who stated that he could have used the support of Drishtee, but "nobody came. I wished that somebody might have come and helped with any problem that I face, but there was nobody to help me." They both appeared to be passive in this regard and placed the expectation on the training provider to approach them first.

The Rangoli shop-keepers had in common the fact that much support was given in the beginning, the pioneering stage of the Rangoli shop, in which they received a start-up package, including a loan of 20 000 Rupees which they have to return, and support with the advertisement. One lady mentioned that Drishtee organized a competition. But it seems that, after this initial phase, Drishtee only kept in touch when it was necessary by supplying the shops with goods. LE4 expressed her desire to receive more specific support from Drishtee, as her income was too low.

Figure 19: Handholding with the training provider

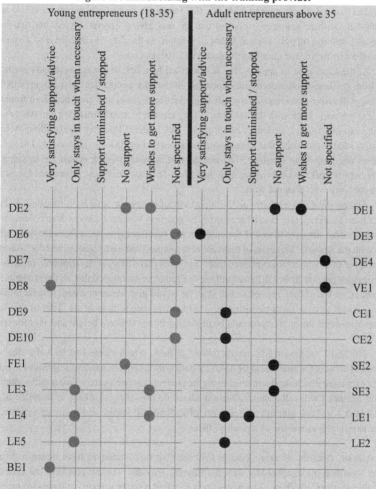

Source: Author's own draft. For the legend regarding the entrepreneurs, see Figure 10 (6.1.1.1).

Now the perspectives of the training managers and trainers regarding the handholding with the trainees will be examined. A Drishtee manager in Madhubani said that, regarding the Rangoli shops, for a period of 2-3 years they were in touch with the Rangoli shop-keepers, "because not only [do] they have to open shops, [but] we have to see that it is sustained also. Because at times they start something. But they will not be able to sustain. So, we are making them sustainable". Another Drishtee manager explained that, regarding the handholding with farmers or dairy farmers, they used to give support afterwards for six months, including doing visits, but this has stopped, because under PMKVY this would not be required anymore. In line with this, another Drishtee manager explained that he was giving general advice, but not more individually tailored support, such as financial support, as the duty of Drishtee is to give their trainees the certificate, and after that the duties of Drishtee end.

At the same time, it appears that the practical handling of the entrepreneurs seems to differ from training location to training location. In the Munger district the manager said that "Drishtee will visit the students afterwards to get feedback about how they benefited from the training. They will check with the students, if they can apply their knowledge, i.e. what is the student's capacity to read the disease of animals." That sounds like active handholding. At least DE8 from Munger confirms this impression, as he pointed out that he felt honoured to get information and help from Drishtee. The manager of another training centre for stitching in Munger, which works in cooperation with the Indian Tobacco Company, promised to send any student for work who does not work.

Among the dairy teachers, different patterns of behaviour have been revealed. A dairy teacher from the Bhagalpur district answered the question whether he keeps handholding with the students after they finish and if the students can receive any support from Drishtee afterwards with the answer: "No, most of them are able to open their own business and do." Another dairy teacher from Bhagalpur explained that "that is not part of my job description". In contrast, a further dairy teacher from a different village in Bhagalpur mentioned that all the students had his number, so that if they wanted to call him, he would give his advice with respect to dairy farming, and if anyone faces problems he would even come to their home. Two of the dairy teachers added that the mobilizer would be in touch with the students before and after the training.

The RSETI manager presented a follow-up form, which means that he had a clear description of what is the true extent of RSETI's duty with respect to handholding with the trainees. He explained that he keeps in touch with the students for two years: "Two months after the training finishes I will call each candidate to check on how they are doing. If someone is not able to work I will ask him to come into my office and will inspire and motivate him to start related to training, as a dairy, as a poultry, fishery."

In conclusion, both the entrepreneurs and the training representatives have drawn a diverse picture regarding the handholding with the training provider. This ranges from the entrepreneurs' perspectives, which includes being "very satisfied with the support" and runs to "missing any support", and from the perspectives of the training providers, which includes the promise of having "provid[ed] active support" to "stopping any support once the training was completed". In this light, it turns out that the entrepreneurs' success in business or the application of training inputs or training evaluation is not necessarily correlated with the handholding of the entrepreneurs with the training provider. To exemplify this, DE2 and DE7 applied specific training inputs and had a medium or high increase of their businesses, and DE7 even evaluated the training course with an A, yet they were both keen to get support from the training provider. However, those who stated that they had received "no support" either did not start the business successfully or graded the training course with a C. On the contrary, those who were very satisfied with the support tended to perform very well or to grade the training course very positively. This leads to the conclusion that the entrepreneurs only partly benefitted from the assistance of the training providers, or in the case of the wealthy dairy farmers from RSETI, the connections of the training provider to the bank. Yet in the important aspects of gaining access to financial support and engaging in social networks, which is crucial for the successful growth of the business, the training providers failed in most of the cases to provide any assistance or to use their connections to banks, farmers' clubs or SHGs. This leads directly to the next point, the engagement in social networks.

6.1.3.3 Engagement in social networks

At this point, attention is next given to tertiary socialization agents, which have a direct impact on the entrepreneurial activity, i.e. cooperatives, farmers' clubs and SHGs. The importance of such institutions for the individuals' ability to gain access to financial support and their development of sustainable livelihoods have been outlined in the theoretical framework. With regard to structural limitations, the question at hand concerns the extent to which the surveyed entrepreneurs benefitted from such institutions as social networks. Have they been able to take advantage and extend their agency? At this time, insights into this issue are gained not only from the entrepreneurs and training representatives, but also from a village elder and a village head.

Insights from the entrepreneurs
Figure 20 illustrates the engagement of the entrepreneurs in social networks. It is differentiated between high or moderate engagement in farmers' clubs, SHGs, politics, Panchayati Raj-institutions[82] or NGOs, or no engagement. For the purposes of obtaining a full picture, some female entrepreneurs from the group interview in the village of Borwah, Bhagalpur district, who mentioned that they were a part of a SHG, have been included as one unit in addition to the single entrepreneurs.

Among the 25 entrepreneurs, only three were actively engaged in a farmers' club or SHG, which raises the question of what motivated some to become engaged and what held others back from it. This will be evaluated on a case-to-case basis. These examples have been found in all the three target-districts. In contrast, the entrepreneurs who did not take advantage of such groups will be introduced with the aim of understanding their motives.

---

[82] A public body of local self-government.

Figure 20: Engagement in social networks

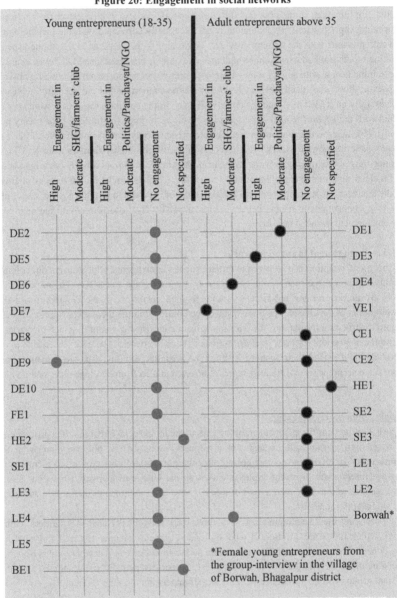

Source: Author's own draft. For the legend regarding the entrepreneurs, see Figure 10 (6.1.1.1).

132

Two entrepreneurs have been interviewed who were both managers of farmers' clubs. One was a dairy entrepreneur from the village of Afzalnagar in the Munger district (DE9), who was the manager of a dairy farmers' club, and the other was the vermicomposting-entrepreneur (VE1) from the Madhubani district, who since 2013 was directing a farmers' club under NABARD. The dairy farmers' club managed by DE9 was managed under Atma (Agricultural Technology Management Agency[83]) and had 20 members. All of them were dairy farmers from the village of Afzalnagar in the Munger district. This group of farmers had only existed for the last six months. But he mentioned that another farmers' club had also existed in the village for the last four years, which was open to any farmer related to agriculture. The only hurdle to enter the dairy farmers' group was a one-time nomination of eleven Rupees.[84] The benefits of joining that group relate to gaining access to both financial and physical capital. According to his information, he was personally granted a loan from the Bihari Government through the dairy farmers' club.[85] However, although the 20 group members currently did not receive any loan, they were planning to apply for it in future. Even someone who does not own any cows yet could become a member and take a loan through the group to purchase cattle. With respect to access to physical capital, as a group they received a bulk of cattle-food from the block-office, which they could easily distribute to the farmers from there. DE9 was further engaged in the ITC Munger Dairy Project, a milk processing plant in Munger, which was a dairy project that collects milk and where he managed all the villages. VE1, the APL-plus entrepreneur from the Madhubani district, was also the coordinator of a farmers' club under NABART, as a part of which he also received his farmers' credit card.

DE4, the APL-plus dairy entrepreneur from the Madhubani district, also mentioned that he was a part of a farmers' club, but that this was not a registered, but only a maintained farmers' club. Two trainees who have been interviewed during their Drishtee dairy training course in Munger mentioned that in their family at least one member was part of an SHG group for dairying. The unit leader of Fateh in Bhagalpur mentioned that there was one SHG in the same village, which was named the Lee Laxmi Self-employment group, but which was not authorized by the Government of Bihar, and that some of the seven ladies trained by her were part of that group. They collected from each member 50 Rupees[86] a month, and if anyone needed support, a loan was granted. In addition, all the female dairy entrepreneurs from the group-interview session in the Borwah village, Bhagalpur district, were part of a women's SHG, which had the same structural set-up. Each member had to pay in 50 Rupees a month, and if they saved 4 000 Rupees[87], the government would approve a loan of 10 000 Rupees.[88]

Among those entrepreneurs who were not engaged in a famer's club or SHG is LE2 who used to be part of a women's SHG in the past. She did not give any reason behind why she stopped attending it. In the case of LE1, other women told her to become part of a SHG, but

---

[83] See under http://www.atmamunger.org/.
[84] Approximately 14 cents.
[85] The increase of his monthly income from 10 000 Rupees to 25 000 Rupees was not directly linked to his function as manager of the dairy farmers' club, but he explicated that this was an additional income through his involvement in the milk production through ATMA and ITC.
[86] Approximately 60 cents.
[87] Approximately 50 Euros.
[88] Approximately 125 Euros.

her husband told her not to engage in it, so she did not engage in it. Meanwhile, DE8 explained that he did not want to join because "I need family, have no time to go outside the village for a meeting, because in SHG group they have monthly or weekly meetings, so I don't want to do this".

Other engagements include DE1, who was a member of the political party BJP, and DE3, who was the village head for two terms.

These examples show that the individuals' engagement in social networks is very crucial for improving one's own structural conditions, as it helps to gain access to credits. Especially people from a poor background can benefit from it, as they hardly get access to credits without being part of a SHG. Thus, only three out of 25 entrepreneurs actively engaged in a farmers' club or SHG. In this respect, contrasting examples could be revealed: in one case all women were part of a SHG, but in most other cases hardly any such membership was registered. It must further be noted that these few examples are certainly not representative of the dairy farmers and other micro-entrepreneurs in rural parts of Bihar in general. Another example from a dairy training course in Munger also revealed that among the trainees ten out of 30 participated in a SHG. Hence, for the surveyed entrepreneurs who were not engaged in a SHG, the question now arises whether this was due to a lack of information about them. Do the training providers inform them about different opportunities to become socially connected as entrepreneurs, and do they provide any assistance in their doing so? Are the training providers in touch with any farmers' clubs, SHGs or cooperatives? This will be elaborated in the following section.

Insights from the training representatives and village elders/heads

During the survey, hardly any examples were found in which the managers of the training centres, or the trainers themselves, said explicitly that they actively assist their graduates to become engaged in social networks, such as a farmers' club or SHG, or that teaching on such a topic was incorporated into the curriculum. On the contrary, one dairy teacher explained clearly that "I do not teach this kind of stuff", but he did mention a mobilizer who would support the trainees personally afterwards. The mobilizer who was interviewed in the Bhagalpur district also did not state that he gave out any advice on how to become engaged in a farmers' club or SHG.

In the Bhagalpur district, two representatives of the village community – one village head in the Tetri village, in which a dairy farming-class of Drishtee was visited, and one village elder in the Murli village, where the charkha workers had been visited – were interviewed in order to obtain further insights regarding the structural conditions of villages. The village head of Tetri estimated that 40% of the farmers in the village were engaged in a SHG under Jeevika.[89] He further mentioned that he personally helped to develop the SHGs, and shared his desire to help young entrepreneurs to obtain a loan if they are denied one. The village elder in the Murli village referred to the strong sense of community in his village, saying that "all of the people of this village collected some money, after that they opened a school". He further added that the government covered only a small share of the costs, while the biggest portion was collected

---

[89] A community-based organization also known as Bihar Rural Livelihoods Promotion Society, in which farmers are organized into SHGs. See also 2.3.2 in the theoretical framework.

by the village people. These examples show the practical benefits of engaging in social networks. They explicate that, through common strength, the limitations imposed by one's structural conditions, such as the lack of basic infrastructure (e.g. a school), can be changed. Likewise, entrepreneurs can overcome many obstacles through engagement in social networks.

In conclusion, it can be noted that, on the one hand, the engagement of the surveyed entrepreneurs in social networks was relatively low, a result which was accompanied for most of them with no or very limited access to financial support. But on the other hand, the entrepreneurs were very family-oriented, and, through their family-background, they usually had a strong identification with their self-employed activity. From this point of view, they also often identified strongly with the village community, e.g. in the case of the charkha or handloom workers and the dairy farmers, since these activities constituted a core element of the respective community. This strong community-mindset gives a ray of hope that for the young entrepreneurs the personal boundaries to become engaged in social networks are, little by little, being dismantled, as the village community performs the function of being a significant socialization agent and holds a lot of potential with respect to developing common strength and tackling structural limitations. From the perspectives of the training providers, it can be concluded that the active support given to push the entrepreneurs to become engaged in SHGs and the like, and to apply for a loan, was very limited. This seems to be a serious omission, as it was shown under the application of training inputs (6.1.2.5) that the majority of the entrepreneurs successfully managed to put their reflexive knowledge, which they gained through the entrepreneurship training, into practice. This leads to the conclusion that the following assumption is also true: if the trainers were to draw more attention to showing how to cope with structural limitations, i.e. how to apply for a loan and how to engage in a farmers' club or SHG, it would also be positively applied into practice. This leads to another key finding of the analysis: the significance of the development of personal initiative in connection with the capacity to cope with structural limitations and to gain access to sustainable livelihoods. This factor will be elaborated in the next section.

6.2   The importance of personal initiative in the development of sustainable livelihoods

In viewing the disadvantaged youth sector within the area of tension between agency and structure, the following relationship was presumed. On the one hand, there are structural conditions, which are limiting or enabling structures, and on the other hand, there are proactive or reactive behaviour patterns. An individual's action can be considered as proactive if limiting structural conditions are being coped with or enabling structures are being taken advantage of. On the contrary, reactive behaviour patterns mean simply to accept limiting structures instead of developing a coping strategy or taking advantage of the available enabling structures. The survey revealed the significance of the development of personal initiative in successfully combating structural limitations, which goes beyond the simple application of training inputs. Personal initiative can be equated with proactive behaviour. In the category system, personal initiative has been defined as any type of proactive behaviour that contributes to running the business better, engaging socially or availing support, which goes beyond one's own routine, one's local tradition, or what has been taught in the entrepreneurship training. In the following section, this factor will be considered – in agreement with the specified research question 2 and its hypothesis – in connection with gaining access to various livelihood assets and the development of proactive as well as reactive behaviour patterns.

Among the surveyed entrepreneurs five major proactive behaviour patterns and five reactive behaviour patterns were identified. Figures 21 and 22 assign both these types of patterns to the young and the adult entrepreneurs (below and above the age of 35). If proactive behaviour patterns are very dominant, they are considered as "very proactive", while if they are less dominant, they are considered as entrepreneurs showing "some proactive behaviour". If no proactive behaviour is visible, they are considered as "being solely reactive". The first proactive behaviour pattern, "proactively developing the business" means to show proactive behaviour with regard to the management of the business activity. This could be exemplified by a significant expansion of the business. The second proactive behaviour pattern refers to "diversification", which, on the one hand, could mean "agricultural diversification", e.g. if more than one type of fruit or vegetable is grown, but also if dairy farming is combined with other types of farming; or, on the other hand, it could refer to a diversified livelihood portfolio with a further income-generating activity, which does not stand in relation to the entrepreneurial activity in which the training was taken. The remaining three proactive behaviour patterns signal a regard for the participation in further training, the engagement in social networks (SHG or farmers' club), or the means of availing access to financial support. On the contrary, reactive behaviour patterns imply a passivity regarding one's structural limitations, which become visible in simply following one's own routine or tradition, in not starting the business at all, in showcasing a reluctant attitude towards taking further training or engaging in social networks, or in demonstrating passivity in applying for a loan or acquiring other means of financial support.

A significant division is discernible between the proactive and the solely reactive entrepreneurs. Each case is considered thoroughly under 6.2.1 (the passive ones) and 6.2.2 (the proactive ones). The two entrepreneurs who were still undertaking the training at the time of interview, namely DE5 and SE1, have been left out of this consideration, even though they were already experienced entrepreneurs, because the main focus is placed in this section on the time since the training. The goal is to make a distinction between the simple application of training inputs and the development of personal initiative.

At this point an important side-note seems appropriate concerning three factors that affect the reliability of the information provided. Firstly, these findings are only based on one interview session with the respective entrepreneur. Therefore, they depend on the respondent's openness and willingness to provide the information, and as such they have no claim of revealing absolute certainty. So for instance, if an entrepreneur has been identified as being "solely reactive", it cannot be completely ruled out that an important aspect may not have been covered. Nevertheless, the information provided gives a coherent picture, which allows one to build contrasting groups of entrepreneurs. Secondly, it seems important to note that, from the degree to which they act proactively or reactively, no implications can be drawn regarding their level of expertise, by which is meant how well they master their profession. Someone who grew up in a traditional handloom family and learned that profession from scraps is without any doubt an expert in his or her field. And yet, such a person might be considered as being "solely reactive", because he or she was solely acting within a certain scope of action, without actively changing the surrounding structural limitations. Thirdly, there are many entrepreneurs who were pursuing or had already attained a University degree. This line of occupation could certainly provide a hint in favour of their showcasing proactive behaviour patterns. Nevertheless, for the analysis only proactive behaviour in relation to the development of self-employed activities is taken into account. An evaluation of proactive behaviour in relation to further educational achievements would have extended too greatly the scope of the survey.

Another aspect, which especially affects the very poor, is the fact that these people survive and manage to make a living under very harsh and restrictive conditions. This should be proof enough that they have a mentality to survive, which means they are still active, even if they are considered as "being solely reactive". In compliance with Giddens' theory of structuration and Klocker's (2007) notion of thin and thick agency, both behaviour patterns are considered as action, being passive and being proactive, as the individuals who are categorized as exhibiting both behaviour patterns have reasons to act in a certain way, e.g. to not apply for a loan. A manager from Drishtee said that, regarding the entrepreneurs from BPL, "they are all survivalists", which means that they are confronted on a daily basis with structural limitations and, hence, must develop coping strategies. Therefore, their reactive behaviour patterns are also part of their coping strategy.

6.2.1 The fate of the passive and reactive entrepreneurs

Figure 21 and 22 show that out of 23 surveyed entrepreneurs, 15 from both an APL and a BPL background have been categorized as being solely reactive. This means that they did not develop personal initiative regarding their structural limitations. First, the solely reactive entrepreneurs who were young, and then those who were older will now be introduced.

The ladies who were trained in making hand-bags, SE2 and SE3, immediately stopped doing so after Fateh stopped providing the material. As a result, they completely depended on the organization and did not develop any kind of personal initiative to deal with the issue. SE3 exclaimed that she decided not to take a loan from the bank because she was afraid not to be able to pay it back. SE2 returned to her previous activity, which was working as a midwife. A similar case is FE1, who completed her training in inland fisheries with RSETI and then fell into difficult circumstances. Yet she did not seek any support from RSETI and did not take any initiative to solve the issue, which prevented her from establishing her business in fishery.

In the case of the three young Rangoli shop-keepers, LE3, LE4 and LE5, who were all recruited by a mobilizer from Drishtee, all seemed to be working with great commitment, in spite of the other duties they may have, such as LE4 who was still attending class twelve in school. Yet nevertheless, they all seemed simply to follow their traditional routine and not to develop any personal initiative regarding the structural limitations. In the case of LE5, it was the family who motivated her to become self-employed.

Figure 21: The development of personal initiative in young entrepreneurs (below the age of 35)

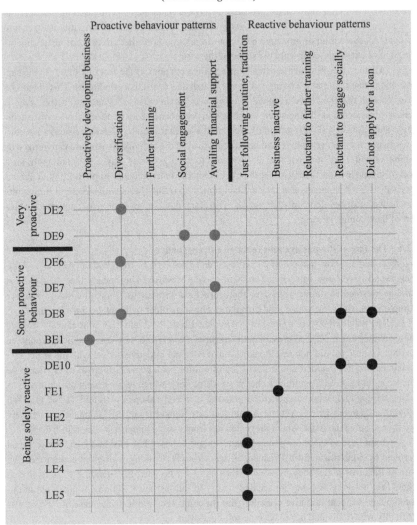

Figure 22: The development of personal initiative in adult entrepreneurs (above the age of 35)

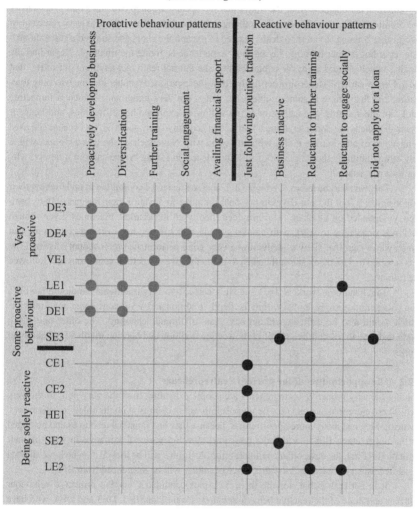

Source: Author's own draft. For the legend regarding the entrepreneurs, see Figure 10 (6.1.1.1).

In considering the young entrepreneurs from APL, the father of DE10, a young female entrepreneur, explained that they currently had no income because the cow was pregnant. They did not wish to expand the dairy business further, as she was still pursuing a BA degree and was planning to attain a government job. Therefore, they also did not apply for a loan.

Now, to look at the older solely reactive entrepreneurs, LE2 from APL, a mother of three children between 15 and 18, whose husband has a customer service point in a bank, had no improvement in her business since the time when she opened the shop two and a half years earlier and despite the fact that she took the training only one year earlier. She was very reluctant to take further training. Even though she wanted to expand the shop, she showed a resistant attitude regarding further training, saying that "experience is giving me training". Regarding the training which she did take, she explained that she did not really get anything out of it: "that much I knew earlier". Regarding her difficulties, she mentioned that her income was only fluctuating and that "there is some problem but that is not very important – which is impacting [me]". In other words, she cannot or is not willing to explicate her difficulties, and hence it seems that she is not doing anything about it. She certainly knew about SHGs because she used to be a part of one, but she quit it without saying why. Nevertheless, she was a woman with a very strong attitude, showing a lot of self-confidence, in saying "if we can run a family, why can't we run a shop?"

The two charkha ladies, CE1 and CE2, also seemed to be committed in undertaking their responsibilities, but did not develop personal initiative to advance their business. They completely depended on Drishtee, who provided them with the charkha machines. Even though CE1 was privileged to work with a solar system, which eased her working conditions, and which was a sign that she was delivering a very good performance (on account of which she was selected by Drishtee), she still behaved within the given structural conditions and followed the routine.

Both handloom workers, HE1 and HE2, came from very traditional handloom families. The younger one simply abided within the family tradition and never considered anything else, which would also be difficult without any years of formal schooling. The older one, HE2, explained that all his children simply followed the tradition and became involved in handloom work.

6.2.2 The opportunities of the proactive entrepreneurs

Among the very proactive entrepreneurs are people extending from the very poor to the very rich. These entrepreneurs can now be purposefully set in contrast with the solely reactive ones. Some of them had many more opportunities, because they had financial resources and belonged to the upper class. But how did the poor ones develop personal initiative in spite of their limitations? First, the three older candidates from APL-plus will be looked at, who have all been classified as very proactive. After this, three examples will be introduced from BPL.

It is noticeable that, for the three APL-plus candidates, all five proactive behaviour patterns and none of the reactive behaviour patterns were identified. DE3 and DE4, who have been trained in dairying by RSETI, were both from the upper classes and invested a lot in training related to dairy farming or agriculture. Besides the RSETI dairy training, DE3 received entrepreneurship training related to agriculture in at least four more locations. He named the Agricultural University in Pusa, Bihar, a place around 100 km south of the training location of Saurath in the Madhubani district, which has affiliated research centres, Kanpur and Lucknow in Uttar Pradesh, and Hyderabad in South India. Furthermore, the bank selected him – in the words of the translator – as "a successful farmer". DE4 also went to Pusa and Lucknow and was described as being a "very progressive farmer" by the translator. Part of the portfolio of DE4 was fishery, growing potatoes and sugar cane. He was experimental, saying that, for example he was doing onions on an experimental basis. He even got a solar system which was subsidized for up to 90% by a bank. VE1 too had been described by the translator as being a "very good progressive farmer – that means a very good farmer, he had also won some prizes".

Furthermore, three individuals from BPL were also classified as being very proactive. DE2 was very proactive in following diversified businesses opportunities to earn his livelihood. The interview took place at his general store, which he was running in addition to his dairy farming business. But he explained that he had done a lot of things before he opened the shop: "I sold newspapers, I had a small kiosk there, a beedi shop[90], so after that I started again this thing." This suggests that he was not necessarily very proactive with regard to dairy farming, but that he was more so in gaining access to livelihoods. In addition to DE2, DE9 stands out as he was the manager of a dairy farmers' club with 20 members. In this role, he was helping other farmers to gain access to financial or physical capital. He was also building strategic partnerships with the milk collection centre. Another example is that of LE1. LE1 managed to expand her small Rangoli shop significantly by adding a beauty parlour to her shop, which is a form of diversification. She explained that she spent all the capital on her own. She also showed personal initiative by talking to some of the agents, who told her: "If you open that shop outside we will give you more products." That indicates clearly that she used connections to gain greater market shares. As a result, she was selling more products than previously. In addition, she set a good example in managing finances and breaking through barriers: "I also returned the loan which I got from Drishtee. I did not take help from my husband in returning the loan. I used the income of the shop to return the loan and after returning the loan I used the house income in purchasing the products and increasing the products." In considering these three very proactive examples of individuals from a BPL-background, it turns out that DE9 and LE1 have in common the fact that they were both very committed to one activity, and that both became creative in expanding it, whereas DE2 might just have been a usual dairy farmer, but had a very proactive and inventive personality, that pushed him rather to do something on the basis of trial-and-error than doing nothing or simply sticking to one thing.

The entrepreneurs with some proactive behaviour are both younger and older. With respect to dairy farming, DE8 from BPL would actually have to be assigned to the group who displayed solely reactive behaviour, as he was reluctant with regard to both engaging in social networks and applying for a loan. He gave many excuses to explain his decision not to become part of a SHG, such as "I need family, have no time to go outside the village for a meeting, because in SHG group they have monthly or weekly meetings, so I don't want to do this". In addition, he never applied for a loan and exclaimed that he would take a loan if the government was ready to give him a loan. This evidences the fact he was not even making an attempt to try to obtain a loan, but simply accepted the status quo. Yet at the same time, his father died, so he carried a lot of responsibility as the father and head of the household, and was engaged in farming as an additional activity.

---

[90] "Beedi" is an Indian type of cigarette.

In the case of DE6 from APL, she has a diversified business portfolio, as she was a home teacher and very committed to becoming a school teacher one day, while at the same time she still wanted to invest in her dairy farming skills. DE7 from BPL, a mother of four children, exclaimed: "If any cow produces less milk, I will sell it and buy another one which produces more milk." She also borrowed money from another villager. BE1 from APL also delivered a very good performance. She was already an active entrepreneur before the training, but after the training her income more than doubled. She explained that "after [taking] the training my confidence filled up", so she was able to use her confidence as part of her human capital to advance her business.

DE1, as the only older entrepreneur, was proactive regarding the management of cows: "What happens if I am not able to rear all the cows, there are lot of cows, I have extra three cows which I have given to somebody to take care ... and they are also selling the milk and the money is coming to me." In this he showcased the fact that he included further people in taking care of the cows. If he relied just on himself, he would not have been able to keep so many cows, and therefore would have a reduced income. But through this own personal initiative he found a way to improve the business. He also used to run a farmers' club in the past.

In conclusion, it can be said that the development of personal initiative is crucial in order to gain access to productive livelihoods, and to cope with structural limitations, which confirms SQ2. Through the analysis the hypothesis of SQ2 could be proven as well. It could be shown that both proactive and reactive behaviour patterns have a significant impact on how productively the reflexive knowledge is applied to one's self-employed activity. For example, LE1 or BE1 managed to apply it productively, and thus optimized their businesses in a proactive way, while the remaining Rangoli shop-keepers, as well as the charkha and handloom workers, simply followed their daily routine, which means that they applied their knowledge very reactively. This also allows us to draw some conclusions regarding the empowerment of the entrepreneurs to cope with their enabling or limiting structures. The category-based analysis, which was divided into the sections devoted to personal, economic and socio-cultural empowerment, revealed a different set of structural constraints for each dimension of empowerment. These are obstacles to success, for which the development of different proactive behaviour patterns is crucial:

- On the level of personal empowerment, personal and family constraints occur due to one's social status, which is often coupled with a lack of education and the necessity faced by individuals to engage in domestic affairs, such as the family business, from a very young age. These constraints, which are deeply rooted in the societal system, certainly cannot simply be removed altogether through personal initiative. Yet, they can be further limited through proactive behaviour, such as seeking further education and training, which would strategically increase one's reflexive knowledge regarding the self-employed activity, and thus they would "thicken" one's agency to develop the business. In addition, they would help one to cope with structural constraints, that lay in other areas of empowerment.

- On the level of economic empowerment, limited access to financial capital is the most restricting factor. This can first and foremost be coped with by availing financial support through engagement in social networks.
- On the level of socio-cultural empowerment, the reluctance to engage in social networks, e.g. a farmers' club or SHG, must be addressed.

The three entrepreneurs from BPL, namely LE1, DE2 and DE9, have shown that it is possible to develop a high degree of personal initiative in order to address these limiting structural constraints, albeit to a limited extent. Nevertheless, stark contrasts have been revealed between the very wealthy candidates, such as DE3, DE4 and VE1 from APL-plus, who without any doubt occupied a much better starting position than the small micro-entrepreneurs, and thus were able to increase their business to an extraordinary degree in comparison with the businesses of the micro-entrepreneurs, which remained on a low level. This leads to the conclusion that, even though SQ2 and its associated hypothesis could be supported by the surveyed entrepreneurs, the entrepreneurs were still in a position of being restrained by their limiting structural conditions. Their development of personal initiative only helped to limit their conditions to a certain degree, as the area of influence in which they can make a difference through action is limited. Hence, other stakeholders also play a crucial role in order to change the underlying structural conditions, such as the training providers, banks, or political actors. Recommendations for these will be considered in the final chapter, when focus is especially given to this study's conclusions.

These findings lay a useful basis for the next part of the analysis, the type-building analysis. In advance of this, special attention must also be drawn to the individuals' awareness of enabling or limiting structures, which will be elaborated in the next section, as well as to their personal motivation and perceptions. This is especially the case with the young entrepreneurs, who stand in an area of tension poised between following the family's tradition and remaining in low-level self-employed activities within a very constrained environment in rural parts of Bihar, or thriving for higher educational goals and career-centred jobs. This tension will be considered further under 6.4.

6.3 The entrepreneurs' awareness of limiting or enabling structures

The entrepreneurs' perception of enabling or limiting structural conditions determines to a great extent the agency of young entrepreneurs. It was assumed that through the entrepreneurship training the reflexive knowledge of the participants increased. This includes both the technical knowledge as well as one's perceptions regarding the structural conditions. In this section, the second of these two aspects is given focus. Hence, it can be assumed that an increased awareness of the enabling or limiting structural conditions would *thicken* the entrepreneur's agency, and thus his or her capability to cope with structural limitations would increase as well. In addition, the awareness of structural limitations is key for the understanding of why a certain livelihood strategy has been developed, since it belongs together with the development of one's own personal motivations and aspirations, as is displayed within the yellow box in Figure 5 (Chapter 3). In what follows here, the awareness of enabling or limiting structures is viewed in a cross-sectional table (Table 7) together with the development of personal initiative, as this awareness has a direct limiting or widening effect on the entrepreneur's agency. Finally, some important general remarks from the APL-plus-candidates regarding the structural conditions are considered.

6.3.1 The entrepreneurs' awareness of limiting structures

The surveyed entrepreneurs had generally very little awareness of limiting structures, or they were not able to name them. When asked about limiting structural conditions with regard to their self-employed activity, most of them only named the immediate financial or physical needs that they were confronted with. For instance, LE4 mentioned that the income was not sufficient, and wished to obtain support from Drishtee. DE1 exclaimed that he had "not all the facilities available" in order to further increase his dairy farming-business, but he did not go into further details about what he thought he lacked specifically. CE2 expressed her discontent about not being able to work on an electric charkha machine, and SE2 explained that the materials for making hand-bags were not sent in the right condition, which is why she, as well as SE3, was not able to continue in this line of work. DE7 is the only entrepreneur, who mentioned personal constraints to pursue further education: "Due to my many children I am unable to. When the children are grown up I will go for more education."

Many entrepreneurs did not name any limitation or exclaimed that everything was fine, such as DE5, who said that the "customers come to me, I don't face any problem to go to the customers and influence them to buy milk from me." In this account he was describing his daily routine, which leads to the conclusion that he had become so acquainted with his situation that he no longer saw anything that needs to be changed. Likewise, LE2, LE3, DE6 and DE8 did not name any constraints. It seems that they too all managed to keep the business running, but they were at the same time confined to their local traditions and did not see things within their areas of action to improve the business. Thus, their business remained on a very low level, which stands in contradiction with the desire to increase the income, as was explicitly expressed by LE3, DE6, and DE8.

Hence, it appears that a correlation exists between one's individual awareness of structural limitations and one's development of personal initiative, which is illustrated in Table 7. While LE2 and LE3 were solely reactive, DE6 and DE8 were only proactive with respect to the diversification of their livelihood portfolio, but not with respect to the development of their business. DE8, for example, was very passive with respect to his social engagement or in applying for a loan. In connection with the previous findings, this leads to the conclusion that one's awareness of limiting or enabling structural conditions is crucial in order to productively apply one's own reflexive knowledge to one's self-employed activity, and in turn, to cope with structural limitations. This supports the hypotheses for SQ1 and SQ2, as the reactive entrepreneurs had little or no awareness of these structural conditions, while the proactive entrepreneurs were much more aware of them, and consequently were able to apply it in a productive way to their self-employed activity. At this point a side-note regarding those entrepreneurs with little or no awareness seems appropriate. These entrepreneurs certainly knew that they were poor and highly constrained. Yet it seems that they also showed a limited awareness regarding their agency, i.e. their own capacity or ability to act in order to change these restraining conditions. This might also be a sign of a lack of entrepreneurial competencies.

Thus, the other entrepreneurs who took advantage of their enabling structures are examined next, in direct contrast to the reactive ones.

6.3.2 The entrepreneurs' awareness of enabling structures

One major characteristic of the proactive entrepreneurs is that they were aware of their enabling structural conditions and used this knowledge strategically in order to advance their self-employed activity. The APL-plus candidates in particular had a very high level of knowledge with respect to enabling structures, and thus behaved proactively with respect to all the identified areas. The difference between them and the proactive entrepreneurs from BPL is that the APL-plus candidates seemed to possess deeper insights into the wider structural conditions of the general system, and thus they gave general commands and advice for small entrepreneurs (see under 6.3.3). Thus, the question at hand is the following: How did the proactive entrepreneurs from BPL perceive their enabling structures? In this regard the following answer can be given. DE9 and LE1 both stand out, due to their awareness of these enabling structures and their development of personal initiative. As a manager of a dairy farmers' club, DE9 was aware not only of his personal needs, but also of the specific needs of the 20 members of the club in terms of their access to financial and physical capital. Thus, he used his knowledge to specifically cater to these needs. Furthermore, he was aware about the further processing of milk before it is packaged in the Bhagalpur collection centre, as he was directly engaged with the milk collection centres ITC and Sudar, and as he also personally benefitted from collecting milk. LE1 also had a broader view, as she did not just take the entrepreneurship training to open a Rangoli shop, but also took a beautification training at a private organization, in order to open her beauty parlour. Furthermore, her awareness of the limited space, especially with regard to the beauty parlour, encouraged her to talk to the agents of the product agencies for the shops, who told her that if she would open the shop in a different location which she can further expand, she would over time be able to accommodate more products. Hence, both DE0 and LE1 were not only aware of their enabling structures, i.e. the dairy farmers' club and further training offers, but they also took personal advantage of it, which for both was accompanied by an increased knowledge. This they effectively applied to their self-employed activities, and thus thickened their agency. This helped them to cope with their structural limitations to a certain degree.

6.3.3 General comments on structural conditions

During the evaluation, the higher level of knowledge and expertise of the APL-plus candidates had been availed by specifically asking them about their perceptions and insights regarding the limiting structural conditions of small-scale farmers from BPL. DE3 highlighted the advantages of dairy farming in comparison to crop farming. For one thing, it is not dependent on seasonality, like the rain season, and therefore income can be generated throughout the whole year. Furthermore, dairy farming is more resilient against natural disasters, such as floods, and thus has the potential to eradicate poverty. He further noted that "small entrepreneurs need to have property as guarantee, which many people do not have, and thus are not eligible for a loan". DE4 also gave the very crucial piece of advice that small farmers should "become organized in a group and do everything in a group", which would foster their development.

With respect to gaining access to financial support, he highlighted weaknesses in the training system, referring to "those people who are able to help them give them training to fill the papers, but after that they forget about them, they do not help them". Furthermore, he expressed his dissatisfaction with the existing training system by saying that "it is not happening properly". When asked what solution he sees in response to this structural deficiency, he replied: "Actually, the government departments are focussing on big farmers, but they should focus on small farmers and after giving training they should be supported." This confirms the findings from the survey of the RSETI-entrepreneurs itself, which have revealed the great contrast between the APL-plus candidates on the one hand, and BPL-candidates, such as FE1, on the other hand. Thus, it confirms the earlier findings that small-scale entrepreneurs from BPL are especially restricted by limiting structures outside of their area of influence, which they cannot resolve solely by developing personal initiative. Nevertheless, the survey has shown that the potential of the disadvantaged youth sector to develop personal initiative, especially to develop their strengths of actively engaging in social networks, should not be underestimated.

Table 7: Cross-sectional table of the awareness of enabling and limiting structural conditions and the development of personal initiative

		Awareness of enabling/limiting structures				
		Aware of enabling structures	Aware of shortcomings in the system	Aware of immediate personal constraints	unaware	No comment
Development of personal initiativ	Being very proactive	DE3 DE4 DE9 VE1	DE3 DE4			
	Some proactive behaviour	LE1		DE7	DE6	BE1
	Being solely reactive			CE2 SE3 LE4 FE1	DE8 LE2 LE3	DE10 CE1 HE1 HE2 SE2 LE5

Not considered: SE1 and DE5.[91]

Source: Author's own draft. For the legend regarding the entrepreneurs, see Figure 10 (6.1.1.1).

6.4 Only Plan B – the mismatch between self-employed activities and career aspirations

The advantage of the qualitative research design is to be able to capture the personal perceptions, motivations and aspirations of the entrepreneurs. In particular, in being ensnared within the area of tension between agency and structure, young people often find themselves poised between the constraints of their family tradition and rural environment on the one hand, and the pull of newly emerging educational and professional opportunities in other parts of the country on the other hand. Therefore, the question matters how the individuals' self-employed activity matches with their personal career aspirations. This is an expedient question regarding the

---

[91] SE1 and DE5 were not considered in the development of personal initiative, as they were still currently taking the training (see under 6.2).

identification of livelihood strategies. This issue was also recognized by representatives of the training providers within the different target regions. A manager of the Drishtee training centre in the Madhubani district said regarding the trainees of the dairy farming course that "cattle rearing is only plan B". In the Munger district, a dairy farming-teacher also estimated that only 50% of the graduates would become self-employed in dairy farming afterwards, while the rest of them would try to get a government job. With regard to handloom workers in the Puraini village, Bhagalpur district, the textile manager of Drishtee pointed to the fact that "many children are leaving the village". These statements underline the assumption that for many of the young entrepreneurs – especially the more educated ones – there is a mismatch between the self-employed activities and their career aspirations, irrespective of the level of accuracy of these statements.

Looking at the young entrepreneurs, a divided picture appears with regard to their personal educational and career aspirations. This is illustrated in Figure 23, which shows on the right side the entrepreneurs' personal aspirations regarding further entrepreneurship training and their personal goals regarding the self-employed activity, and on the left side their educational attainments and career aspirations. What stands out in particular are great differences between entrepreneurs with an APL- and BPL-background. The four APLers pursued entirely or have already completed a Bachelor's degree at university. Among the ten BPLers, only three pursued entirely or have already completed the higher secondary school qualification, and two of them have obtained a Bachelor's degree. The entrepreneurs can broadly be divided into three groups: those who were solely focused on the entrepreneurial activity, those who seemed to be split in their aspirations and tried to pursue both the self-employed activity and a professional career, and, finally, a few who seemed only to want to escape the self-employed activity and solely to focus on their professional careers.

Nearly half of the young entrepreneurs were solely focussed on the self-employed activity, namely HE2, LE3, LE4, DE7, DE8 and DE9. They did not mention any career aspirations apart from their self-employed activity and did not pursue or complete a higher educational degree. Yet, stark differences appear if one takes the levels of these candidates' specification into account, as only very few were more specific about their personal aspirations regarding the self-employed activity. At the one extreme is the traditional handloom worker from the Puraini village in the Bhagalpur district, namely HE2, who did not communicate any aspirations regarding his personal development or the self-employed activity. Then, there are three entrepreneurs, DE7, DE8 and LE3, who simply stated that they were aiming at increasing their business. Two of them were also willing to undertake more training related to their business, about which, however, they did not further specify. For example, LE3 said, "I want to take such type of training which would improve my business." Only two of the entrepreneurs, DE9 and LE4, were more specific and postulated a concrete idea of how they wanted to use further entrepreneurship training strategically to further advance their business. Moreover, LE4 from the Bhagalpur district mentioned that she wants to take training in beautification and open a beauty parlour, which would be an expansion of the Rangoli shop. This was successfully realized by LE1, a Rangoli shop-keeper from the Madhubani district, who also took a training in beautification. DE9, the leader of the dairy farmers' club in the Munger district, stated that he wants to take training to learn "how to make things from milk, like cheese, butter and panier."

Figure 23: Personal aspirations of the young entrepreneurs regarding education and training, self-employment and career jobs

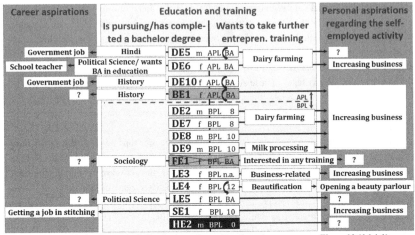

Source: Author's own draft. For the legend regarding the entrepreneurs, see Figure 10 (6.1.1.1).

In addition, a large group of entrepreneurs, i.e. SE1, LE5, BE1, DE2, DE5 and DE6, had personal aspirations regarding both the self-employed activity and other activities, some of whom also thrived for a high prestige job. This is also a heterogenous group, of which two had only completed the school to the eighth and tenth class, while four attained or were pursuing a Bachelor's degree. Among the lower-educated ones is SE1, the one who was currently taking the training in stitching but who was already self-employed with a tailoring shop for many years. On the one hand, she said that "I want to develop my business of stitching", but on the other hand, she wanted the certificate of the stitching course in order to apply for a job, which she explained by the fact that "the rate is not very good in the village, so I am unable to meet my daily necessities". DE2 had also already built up a diversified business portfolio. Besides dairy farming, he was running a general store. But still, he was open to take further training related to dairy farming, and said that "now I want to apply for a loan so that I can expand". In considering the higher-educated group, it stands out that all of them picked a subject in the humanities, which was altogether unconnected with the self-employed activity. Even though she attained a BA in political science, with regard to her personal aspirations LE5 seemed very disoriented on both sides. She had no vision concerning the Rangoli shop, and she also did not specify how she wants to use her BA in political science. BE1, on the contrary, also did not specify how she wants to use her BA in History, but she was very goal-oriented in developing her beauty parlour. DE5 and DE6 were split regarding their personal aspirations. They both wanted to take further training related to dairy farming in order to increase their businesses. At the same time, DE6, who had already completed her BA in political science, was aimed at becoming an elementary school teacher, for which she needed a BA in education. In this, she was being very ambitious, as she was already working as a home-teacher. When asked about

148

her motivation for getting a BA in political science, she answered that she wanted "to get knowledge about political parties in India and all over the world". DE5 said that "if I get the opportunity, I would go and work for another business", and added that he preferably would like to work for a governmental organization.

Finally, there is DE10, who can be grouped with those who did not want to pursue the business further and thrive for a high prestige job. When asked about his future plans regarding the dairy farming business, her father answered that "they do not want to explore this business more, but after getting a BA degree she wants to apply for a government job". The young lady, FE1, who received her training in inland fisheries, cannot really be assigned to one of these groups, as she was currently not engaged in a self-employed activity. Even though she had a BA in Sociology, she did not specify any career aspiration. She only mentioned that she would be interested in undertaking a dairy farming training, as they were keeping a cow in the family.

In conclusion, it can be said that the educational achievement, which is obviously connected to the APL/BPL-status, is a distinguishing feature of the young entrepreneurs, which leads many to feel uncertain and locked between the self-employed activity and their career aspirations. For them the hypothesis proves to be true that, in considering personal aspirations, there is a mismatch regarding these two sides. At least three entrepreneurs seemed to distinctively be following a Plan A, i.e. by getting a government job or becoming a teacher. For them the self-employed activity seemed only to be a Plan B, whereas in the case of dairy farming, the self-employed activity does not necessarily stand in contradiction to other personal goals, such as becoming a teacher in the same village, as the whole family usually helps out in the business activity. Only in the case of the stitching entrepreneur was the aspiration to get a paid job clearly coherent with the entrepreneurship training.

The assumption that there is a mismatch between the self-employed activity and the personal educational and professional aspirations is also supported by the 16 trainees, who were interviewed at various training locations of Drishtee, eight in the area of dairy farming and eight in the area of stitching (Figure 24). Altogether, the share of APL-students was very high; many trainees with an APL-background had been interviewed. It is remarkable that more than half of the trainees had completed or were pursuing a University degree, ranging from a BA to a PhD. Out of the ten of them, four combined their studies with a self-employed activity, and thus they strategically selected subjects such as commerce, economics and sciences. For example, a male dairy farming-trainee who studied economics said: "I want to learn more about the economy and to know how to improve my own decisions for my animals." In contrast, the six remaining trainees at BA-level, plus two of those who had reached class twelve at school, planned to get a high prestige job, but at the same time they kept open whether they also wanted to engage in the self-employed activity. Five of them wanted to get a "government job", of which three clearly cited it as being the first priority, and the self-employed activity only as the second priority. The remaining three trainees were uncertain about what they wanted to do with their degree; for example, a lady from a dairy farming class stated: "I want to become highly qualified as a very known person for my family and the society." Hence, like the young entrepreneurs, the majority of the trainees with a higher educational degree aspired – sooner or later – to get a high prestige job, in areas other than the self-employed activity. The majority selected – also like the young entrepreneurs – subjects in the humanities, such as History. Yet the four trainees who combined their studies with the self-employed activity represent a new group, which had not been counted among the surveyed entrepreneurs. Their livelihood strategy could be described as *business professionalization*, a strategy which leads to the question whether this would have a synergistic effect on their self-employed activity. Would it further widen their agency and capacity to expand their business and cope with limiting structural conditions? The fact that all of these "combiners" came from an APL-background prepares the ground for a further discussion, namely whether this is a serious opportunity for BPLers to raise their self-employed activity to a higher level, and thus to significantly change their limiting structural conditions. This question will be further elaborated in the final chapter of the dissertation.

The trainees with lower educational degrees counted entirely on the self-employed activity for their future, except for two stitching-trainees from the Drishtee/ITC in the Munger district. They came from very poor backgrounds and used to work in the making of incense-sticks, were being retrained in stitching, and wanted to become employed at the DRAP centre of Drishtee. At the same time, they all came – with one exception –from a BPL-background. Thus, they seemed to have no viable alternative, for otherwise they would have also opted for high prestige jobs.

This leads to the following conclusion regarding the surveyed entrepreneurs who are solely being targeted on the self-employed activity. The hypothesis that the self-employed activity is only a Plan B is not automatically refuted by the fact that they did not specify any other personal aspiration. Only very few carried a high conviction that they would not consider anything else, namely HE2 and DE9. For the other ones who were not as determined as these two, it seems true to say that they were simply open to do any type of work which improves their income, on the grounds that "increasing the business", which basically means "improving the income", was, for most of them, the primary motivation to engage in the self-employed activity. The type of activity by which to achieve this goal seemed secondary. This was also proved through the group interview with the dairy entrepreneurs in the Borwah village, Bhagalpur district. In addition to dairy farming, many of them worked as labourers. Those who did not complete higher secondary education said that they wanted to catch up on this.

Figure 24: Future aspirations and educational achievements of the trainees

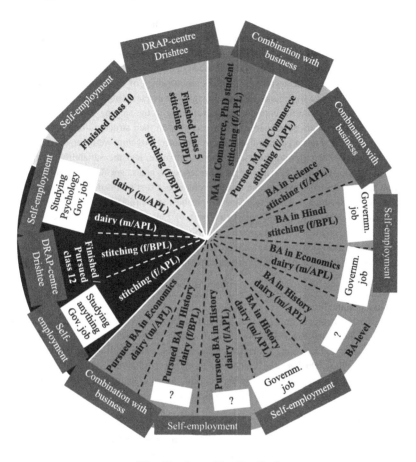

PhD ▪ MA or above ▪ BA ▪ 12 ▪ 10 ▪ 5

Source: Author's own draft.

Part II – The type-building analysis

The second part of the general research question, which interrogates how the young entrepreneurs cope with the structural conditions that enable or constrain them, is a guiding question for the type-building analysis. In the first part of the analysis, the entrepreneurship training course was the focus as the object of research, in order to be able to draw conclusions regarding its function as an instrument for the empowerment of the disadvantaged youth to improve their productive livelihoods through engagement in a self-employed activity. Now, the main focus is shifted from the entrepreneurship training to the varying livelihood strategies, which these entrepreneurs developed in order to cope with their structural limitations. In this way, the previous findings regarding the different dimensions of empowerment, the application of training inputs, their development of personal initiative, and the suitability of the self-employed activity for their educational and professional goals, are the basis for the type-building analysis. In what follows, the case summaries are first outlined on the basis of these previous findings, which will help to define the attribute space (6.5). After this, the number of types is defined and each case is assigned to a type (6.6). Finally, the relationships between each type are analysed and discussed in connection with the specified research question 6, namely what livelihood strategies the young entrepreneurs develop in relation to their educational and social status, in order to cope with the structural conditions that enable or constrain them (6.7 and 6.8).

6.5 Case summaries

The next analysis step is trans-categorical, and relates each of the previously considered categories to individual cases. The objective of writing these case summaries is to compare contrasting cases and carve out differences between them. This will lay the foundations for identifying the types, which are the final outcomes of the type-building analysis. Each type will be defined by its attribute space, following the principle of creating types which differ from each other as much as possible, and where the individual cases of a type-group are as similar as possible.

In writing the case summaries, six contrasting cases have been deliberately selected on the basis of the following attributes. With respect to the grouping of the entrepreneurs, they represent "young adults" – of which some are "youth in transition" – and "adults", and range from BPL to APL-plus. This selection was primarily determined by two key categories: the development of the entrepreneurs' personal initiative and their personal educational and career aspirations. The first of these two categories divides the selected entrepreneurs into two extremes, namely "very proactive" candidates, and the "solely reactive" candidates. A more detailed look also reveals stark differences within each of these two extremes. The latter covers the whole range from "being completely focused on developing the business" to "being completely focussed on a professional career". To pick two examples, DE9 and DE3 represent two entrepreneurs who very proactively developed their businesses, but who had very different starting positions. By contrast, LE3 represents an example of a solely reactive entrepreneur, who simply followed the daily routine. DE10 and DE2 represent the educational escape strategy and the diversification strategy. Finally, FE1 has been chosen by way of an example of an inactive entrepreneur. A headline has been assigned to characterize each case.

Figure 25 introduces each of these cases by linking them with the theoretical foundations that outline how young entrepreneurs developed a livelihood strategy within an area of tension between agency and structure. This was illustrated in Figure 5 with three circles: the red circle considered from an action-oriented perspective how the youth sector gains access to sustainable livelihoods, the blue circle represented the structural set-up and considered the impact of different agents of socialization, and the yellow circle considered the personal motivations, aspirations and perceptions of the young people involved. Likewise, Figure 25 exemplifies the development of livelihood strategies by linking each case with these circles. The blue circle is thus reduced to the four key features of the support structure that have been found to be most relevant, namely family support, gaining access to financial support, engaging in social networks, and handholding with the training provider. Conversely, the red circle considers the self-employed activity in connection with further activities, e.g. pursuing a BA, engaging in other self-employed activities, or assuming the responsibility of being a mother. The yellow circle resumes the educational and career aspirations, which have been illustrated in Figure 23.

DE9 – Intensifying the business through social engagement
Figure 25a:[92]

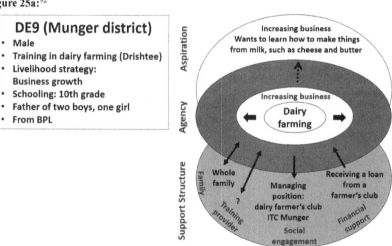

DE9 exemplifies a very proactive young dairy farmer from a BPL-background, who completed his dairy training five month prior to the interview. He was well connected with the support structure. Dairy farming was his main business, and was the main business of his whole family. He was the manager of a dairy farmers' club with 20 members, which was linked to ATMA (see

---

[92] All images designed for the case summaries are subsumed under Figure 25, and are based on the author's own draft.

under 6.1.3.3), and through which he gained access to a loan from a bank. He was also engaged in the ITC Munger dairy project, a milk-processing plant in Munger, where he managed the milk collection in the villages. Through these engagements, he managed to increase his monthly income from 10 000 Rupees to 25 000 Rupees[93] per month, although his quantity of cattle before and after the training remained stable at four.

It can be concluded that this entrepreneur certainly followed a business growth strategy, as he was destined to grow not only his personal cattle business, which he wanted to increase (and which he had indeed already grown by building a bigger stall for his cattle and by successfully applying for a loan), but also his business network, i.e. in managing the farmers' club and holding a managing position at the ITC Munger. Finally, he was committed to increasing his level of expertise and was willing to take further training in milk-processing activities, such as making cheese or butter. This shows that he was obviously thinking ahead and thinking outside the box. In conclusion, this entrepreneur evidently knew how to use his enabling structures to take advantage of the structural conditions to increase his agency.

DE3: Thriving as a successful farmer and expanding through the professionalization of his business
Figure 25b:

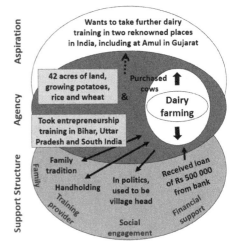

In contrast to DE9, DE3 has been included as an adult entrepreneur in his fifties. He exemplifies someone who is very progressive from the wealthiest section of the society, belonging to the Brahmin-class. As a successful farmer he owned 42 acres of land, where he grew potatoes, rice and wheat, and he increased the number of cattle after the training to 15, from which he received around 90 litres of milk per day. Within his family, farming had been a long tradition. He

---

[93]   Approximately 125 to 230 Euros.

engaged in dairy farming for almost the last 20 years. He was well connected with the support structure, and claimed that he was always in touch with RSETI, where he took his last training in dairy farming about four months ago. He engaged in politics and used to be the village head for two terms. From the bank he received a loan of 500 000 Rupees[94] after taking the RSETI-course. From this course he purchased more cows. In fact, the bank selected him for the loan and requested him to take the dairy training at RSETI as a requirement of his receiving the loan.

Concerning his portfolio specifically, he certainly followed the livelihood strategy, which Scoones (1998) identified as "livelihood diversification". With respect to dairy farming, he followed a business growth strategy through the professionalization of his business. This becomes obvious when viewed together with all the training related to farming and dairy farming that he had already received. He named various renowned training locations of different training providers in Bihar, Uttar Pradesh and Hyderabad in South India. And he still wanted to take more training in dairy farming at two other locations, one at the headquarters of Amul in Gujarat.

DE10: Escaping the self-employed activity and counting on the University for a career
Figure 25c:

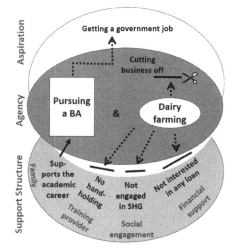

DE10 exemplifies a growing number of young people from rural parts of Bihar, who are pursuing a higher educational degree and are destined to obtain a high-prestige job. Coming from an APL-background, she won for herself the opportunity to pursue a Bachelor's degree. Concerning her self-employed activity of dairy farming, where she conducted a training and which she carried out with the support of the whole family, it seems that she was currently in the process of completely stepping away from this activity, wanting instead only to focus on

---

[94] Approximately 6 370 Euros.

her University studies. In this regard, she was very privileged to be able to do so, as many young entrepreneurs from poor backgrounds are kept back from focussing on higher education due to their specific self-employed activity. At the time of interview, they only had one cow as a family, which became pregnant, so they put the dairy farming activity on ice, deciding not to advance the business any further. Hence, she was very disconnected with regard to the support structure, as is illustrated by the gap between the red and the blue circle: she was not interested in getting any loan, she was not engaged in a farmers' club or SHG, and there was no handholding with the training provider. Her career aspiration was to get a government job. In conclusion, she followed a sort of education escape strategy, which had been adjusted from Scoones' (1998) migration strategy[95]: instead of increasing the business, she was planning to escape her self-employed activity by cutting it off, and to attain a high-prestige government job with her BA in History. For this life decision she had the full support of her family.

DE2: Creating a diverse livelihood portfolio with a low educational level
Figure 25d:

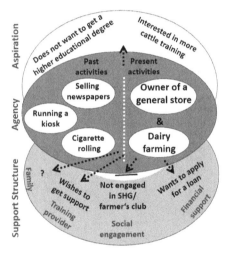

DE2 is an example of many of the young entrepreneurs from rural parts of Bihar with a BPL-status and a low educational degree. In order to survive, he built up a diverse livelihood portfolio, in which dairy farming, which was the field in which he took the entrepreneurship training, was only one pillar. Besides this, he was currently running a small general store, which was only a further addition in a line of various previous low-income engagements, including cigarette rolling, running a kiosk and selling newspapers. With respect to the support structure,

---

[95] The term "migration strategy" seemed inappropriate to use, as she did not explicitly claim that she was planning to migrate. This term would be too abstract, as it does not take into account educational achievements, but only describes the individual's decision to migrate to another location in order to look for other sources of livelihoods.

he had no handholding with the training provider, but wished to receive support from it. He did not engage in any farmers' club or SHG, and did not receive any loan from a bank, but wanted to apply for one. This means that, in spite of his other activities, he was still willing to grow his dairy-farming business. He also noted that he was planning to purchase more cows, for which he needed a loan from a bank. His strong level of entrepreneurial engagement suggests that his family was also supportive of him in dairy farming, although he did not explicitly mention this.

With regard to his future aspirations, he wanted to take more training in cattle rearing. Yet he was not willing to invest in getting a higher educational degree. In conclusion, he followed a livelihood diversification strategy, which means that he built his business on a diverse livelihood portfolio, in which he relied on more than one business activity which was not related to any of the others. As for this entrepreneur's limiting structural conditions, his strategy worked to advance his agency, yet it is not a very conducive strategy for business growth. If he fails with something, he searches for new opportunities, which characterizes him as an inventive survivalist; it is for this reason that he was marked as being very proactive.

LE3: Keeping the status quo by following the daily routine
Figure 25e:

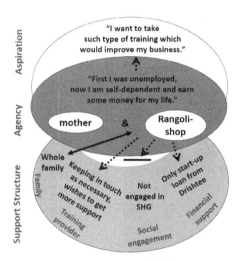

LE3 was a young mother from the poorer section of the society of rural Bihar. She was running a home-based business, a Rangoli shop, with the support of the whole family. Coming from BPL, they lived on very limited financial recourses. Her father worked as a simple labourer and her husband was a farmer. Being located in a small village, she did not receive many customers and made only 500 Rupees[96] per month with the shop, an amount which could more than double during festive seasons. Yet her confidence grew through this business activity, which she

---

[96] Between six and seven Euros.

emphasized by saying: "First, I was unemployed, now I am self-dependent and earn some money for my life." The start-up package that she received from Drishtee to open the shop included a loan of 20 000 Rupees.[97] In addition, her handholding with the training provider ensured the regular supply of the shop with goods, yet she wished to obtain more support from the training provider. She was not part of any SHG and had no access to any additional financial support. As for her development of sustainable livelihoods, she kept the status quo by following the daily routine. She wanted to improve the business, but her behaviour was completely reactive, which means that she did not develop any personal initiative with regard to her limiting structural conditions. She only noted that she wished to take "such [a] type of training, which would improve my business". Therefore, her livelihood strategy is identified as "business preservation", as she managed to keep the business running at a minimum scale without taking steps to develop it further.

FE1: Blocked by limiting structures without having any future aspirations
Figure 25f:

FE1 has been included as an example of a young lady who took an entrepreneurship training, but did not manage to utilize it and get started in a business. She took the training in inland fisheries with the clear goal of becoming self-employed in this area, but eventually she did not get this project started, due to various circumstances. She explained it by the scarce availability of fish ponds. Coming from a BPL-background, she is exemplary as a young woman who is blocked by her limited structural conditions. With respect to her support structure, she was completely disconnected, since she was not part of a SHG, she did not receive any loan from a bank, even though she tried to get a loan, and she did not have any handholding with the RSETI-

---

[97] Approximately 250 Euros.

institution, where she took the training, although they promised to ensure handholding afterwards. She can be marked as being solely reactive, as it seems that she accepted the status quo, rather than developing any personal initiative to avail financial support, which seemed to be her main stumbling block. For this reason, it must be noted that she had responsibilities as a mother of a two-year old child. At the same time, she had already completed a BA in Sociology. Yet with respect to her future aspirations, she was very unspecific, being unable to define any clear career aspirations for herself or outlining what she wanted to achieve when it came to becoming self-employed. She only noted that she would be interested in taking further entrepreneurship training of any kind. In spite of this, she did not give the impression of being disillusioned. Through her family, which was engaged in farming and dairying, she certainly had a stable support system. Yet, it seems that she was not actively supported by her family in getting started with her self-employment in fishery. It can be concluded, therefore, that she did not follow any personal livelihood strategy, but simply adjusted to the structural conditions, which means that she ended up taking whatever was available.

6.6 Defining the types based on the development of livelihood strategies

6.6.1 Determining the attribute space

The case summaries that are individually detailed above have been intentionally chosen in such a way that each livelihood strategy, which was adopted among the 25 surveyed entrepreneurs, would be represented. Before the attribute space is now determined, the question at hand is why livelihood strategies are suitable as a basis for the type-building analysis. The first part of my analysis in this chapter – the evaluative qualitative analysis – has revealed that, through the entrepreneurship training, many of the young entrepreneurs were able to increase their access to productive livelihoods, but often to a very limited degree within a certain structural framework. Yet, only very few entrepreneurs actually managed to change their limiting structural conditions in order to increase their agency. In looking closer at these differences, it turned out that they developed different livelihood strategies in order to cope with their common structural limitations, depending on the interplay of agency, structure and personal motivations and aspirations. Therefore, my next step lies in utilising these analysed strategies to carve out the differences and similarities of the entrepreneurs, and to group them in such a way that conclusions can be drawn regarding how and why they cope with their limiting structures in a certain way.

In order to determine this attribute space, the question occurs of how many attributes to consider. For reasons of clarity and comprehensibility, it seems feasible to visualize the livelihood strategies within a matrix which has both a horizontal and a vertical dimension. For this, two key attributes are needed that are most appropriate to represent different livelihood strategies. In looking at the case summaries, it appears that livelihood strategies have been developed in relation to the following core themes:

- Firstly, regarding the development of personal initiative, which ranges from being very proactive to being solely reactive. More proactive behaviour patterns resulted in the following livelihood strategies: business growth, business growth through professionalization, and livelihood diversification (which for some meant agricultural diversifycation, and for others the diversification of unrelated businesses). On the other hand, the more passive-reactive behaviour patterns resulted in the education escape strategy or the business preservation strategy.

- Secondly, regarding the degree of the individual's determination in relation with his or her personal aspirations. These range from being solely focussed on self-employment – which is the case with the strategies of preservation and optimization or the diversification strategy, e.g. through livelihood diversification or by incorporating a new venture within an existing business, e.g. by enhancing the Rangoli shop with a beauty parlour – to being focussed on multiple disconnected activities at the same time. This latter focus would also qualify as a form of livelihood diversification. Also to be included is a focus on attaining a higher educational degree and getting a high-prestige job, i.e. the education escape strategy. In addition, two of the entrepreneurs with an inactive business have been recognized as "undetermined", as they did not seem to follow any specific livelihood strategy, and, therefore, they would not really match in any of the other categories.

The identification of these features seems sufficiently vital for the determination of the attribute space. Of course, further related issues could also be considered for the grouping of the entrepreneurs, such as their level of focus on personal empowerment, which is reflected in their personal investment in education and training, e.g. in pursuing a Bachelor's degree. Yet, in considering the development of personal initiative and personal aspirations, it has been shown that the entrepreneurs with higher educational degrees are not a homogenous group, and that, therefore, it would not be very meaningful to simply merge them together as career planners. Furthermore, other dimensions of empowerment, such as the individuals' levels of economic and socio-cultural empowerment, could be considered as another dimension for the attribute space. However, these are complex dimensions that cannot be easily quantified. Moreover, it would be too narrow to consider them by using individual parameters, such as "increasing the business" or "engaging in social networks", as not everyone who was able to increase their business through the application of training inputs could be considered as economically empowered. Furthermore, those who were not part of any SHG or farmers' club could not automatically be labelled as not being socially empowered. On the contrary, consideration of the development of personal initiative is a precise attribute to distinguish the entrepreneurs, as it is a factor that looks across categories which are related to both the economic and the sociocultural empowerment of the entrepreneurs. Therefore, the two core issues mentioned above, the development of personal initiative as well as the individuals' degree of determination in relation to their personal aspirations, are chosen as the two key attributes for the type-building analysis.

6.6.2 Definition of types
According to the attribute space that has been defined above, it makes sense to narrow the identified livelihood strategies down to four core strategies, on the basis of which the following four types, plus one additional type for the undetermined entrepreneurs, have been created.

These are illustrated in Figure 26 in the form of a matrix. For the type-building analysis, all surveyed entrepreneurs, who have been interviewed individually (i.e. both youth and adults), are considered. In this way, differences regarding the development of livelihood strategies and portfolios can be identified between the older and younger entrepreneurs, from which recommenddations can be derived for the younger ones, with regard to both what they can learn from the older entrepreneurs and what they need to do differently when it comes to dealing with their limiting and enabling structural conditions.

In Figure 26 presented below, the youth are placed in the blue middle section, while the adults are placed within the upper and lower edges. Along the horizontal axis is placed the degree of determination, which ranges from being undetermined to being solely focussed on self-employment, and from this to being focussed on multiple activities. Along the vertical axis is placed the development of personal initiative, which, again for reasons of simplicity, only has the distinction between reactive and proactive. In accordance with these demarcations, it is now possible to align the livelihood strategies to the following core strategies, with their respective type-names:

- The preservers follow a preservation strategy. They are solely focussed on the entrepreneurial activity and are solely reactive with regard to coping with structural limitations.
- The career planners follow an educational escape strategy. They have a diverted focus by being focussed on pursuing a higher educational degree and getting a high-prestige job. At the same time, they are solely reactive when it comes to their development of the self-employed activity. They are considered as reactive, even though they may also have developed personal initiative with regard to their studies. Yet at this point, the development of personal initiative is limited to income-generating activities, which have a widening effect on their agency. In the case of the career planners, they have not all completed their education yet, and therefore their desired high-prestige job remains only an aspiration. With respect to the entrepreneurial activity, they were solely reactive.
- The optimizers are either mono-structured optimizers, which means that they focus on the optimization of one self-employed activity, e.g. dairy farming, or they are diversified optimizers, which mean that they aim at the optimization of their diverse livelihood portfolio. In other words, they are either solely focussed on one self-employed activity or on a diversified set of activities. At the same time, they were proactively developing the business. The diversified optimizers used diversification as a business growth strategy, e.g. through growing multiple crops in combination with dairy farming. The reason for assigning these adult entrepreneurs to the optimizers, instead of the diversifiers, is to place special emphasis on the fact that they developed personal initiative when it came to the optimization of their business of training, e.g. by professionalizing their dairy-farming activity through advanced training, by using new technologies, or by adding a new venture to an existing business. Furthermore, as adult entrepreneurs they have only been included as a group of comparison for the young entrepreneurs. In this respect, they showed more similarities to the optimizers than the diversifiers.

- The diversifiers follow a diversification strategy. Yet in contrast with the diversified optimizers, they were only proactive in focussing on multiple income-generating activities, and not when it came to increasing the business of training. Furthermore, three of them were pursuing disconnected activities, e.g. dairy farming and teaching or dairy farming and running a general store, whereas the diversified optimizers followed an integrated business growth strategy, in which they strategically added one new activity in combination with another.

- The conformists do not follow any definable livelihood strategy. They are only conformists with respect to their structural conditions, which means being solely reactive and undetermined with respect to their personal aspirations.

- It seemed not necessary to create a type for the hypothetical case of being undetermined and proactive.

Figure 26: Type-building analysis based on the development of livelihood strategies

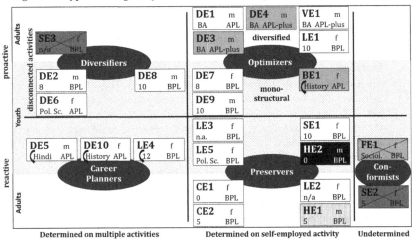

Source: The author's own draft. For the legend, see Figure 10 (6.1.1.1). For the young entrepreneurs who pursued or completed a BA, the subject was added.

It must be noted that, in considering the household portfolios (Figure 18), more entrepreneurs would have to be assigned to the diversifiers. For instance, the husband of DE7 was a carpenter, and in the cases of DE8, DE10, LE3 and FE1, farming was also a part of the household portfolio. Therefore, the following must be kept in mind. The primary focus lies upon the activity in which the entrepreneurs received the training. Only if they personally engaged in further activities as the main entrepreneur were they considered as diversifiers; e.g. in the case of DE10, farming was a part of the household portfolio, yet she was not carrying any responsibility in this domain. The only exception to this trend is DE8, who engaged as a main entrepreneur in both dairy farming and farming, and therefore is considered as a diversifier. Yet he does not match with the optimizers, as he had very passive behaviour patterns when it came to availing the support structure, and he is not considered as a *disconnected diversifier*, since dairy farming is a type of farming.

6.7 Type descriptions

In the following section, each type will be considered in accordance with the specified research question, namely the question, "Which strategy do the entrepreneurs develop with respect to their occupational group, age, gender, social background and educational status?". The goal here is to identify the common patterns of the entrepreneurs within a specific type, so that the types can be clearly demarcated from each other. Furthermore, in considering Figures 5 and 25, it will be asked how the red, blue, and yellow circle are connected: What livelihood strategies do they develop within the area of tension between agency and structure and in regard to their personal aspirations, in order to cope with structural conditions, that enable or constrain them? In connection with the hypothesis associated with SQ7, particular attention will be drawn to the effects of the respective livelihood strategy on the entrepreneur's agency. If the livelihood strategy has an effect, how is it thickening? These questions are elaborated for each strategy as the last step of the analysis. In following this approach, the final specified research question must also be brought into consideration, namely: What recommendations can be derived for the providers of entrepreneurship training with regard to the improvement of structural conditions for young entrepreneurs from socially deprived backgrounds?

In the following, each type is described in detail, explaining in the case of each entrepreneur why she or he was assigned to a specific type.

6.7.1 The optimizers: their characteristic motivation for success

It turned out that in this category, all the young entrepreneurs (aged 18 to 35) were mono-structured, which means that they proactively focused on the optimization of the core self-employed activity in which they carried out the entrepreneurship training, by following a business growth strategy. In contrast, the adult entrepreneurs aimed at optimizing their diversified livelihood portfolio. They could have been grouped with the disconnected diversifiers, but this was not done on purpose. They followed a business growth strategy with respect to the business of training, and, therefore, they needed to be grouped together with the young optimizers. In this phase, first the young entrepreneurs are considered, followed by the adult entrepreneurs.

The four young business optimizers all have in common the fact that they were solely focussed on the development of their business of training, with the ultimate goal of increasing it. In this respect, they showed personal initiative, and two of them also strategically used the support structure. DE7 showed some proactive behaviour in availing financial support, as she borrowed money from someone in the village. Furthermore, she had a high increase of income since the training, despite being a mother of four children between the ages of one and seven. She appeared to be a self-confident woman, saying "if any cow produces less milk, I will sell it and buy another one which produces more milk". This proves her standing as an entrepreneur who is committed at increasing her dairy farming business. In addition, she was open to undertaking more training related to dairy farming. DE9, by the same token, had already been considered in the case summaries as a very proactive entrepreneur with respect to his social engagement by leading a dairy farmers' club. BE1 had a high increase of income since the completion of the training and had very progressively developed the business of running a beauty parlour. She was also pursuing a Bachelor's degree in History, so technically, she could have been assigned to the career planners, yet with respect to her future aspirations, she did not seek to thrive for any high-prestige job, but wanted to further develop her beauty parlour, which is why she is considered as a business optimizer.

In conclusion, it can be noted that, for these entrepreneurs, the red, blue and yellow circles (which reflect the interplay of agency, structure and personal aspirations) are brought into harmony with each other around the goal of increasing the business, which identifies them as true "entrepreneurs" on the understanding of Carland et al. (2007; compare 2.2.3.1) and distinguishes them from the other types. This means that they were empowered to cope with structural limitations, yet only to a certain degree, as their businesses still remained relatively small and limited, although most of them had a high increase of income since the training. The fact that all these entrepreneurs were mono-structured also raises the question of the hypothesis for SQ7, namely whether their agency really "thickened" as well. In the case of DE9, this was certainly the case, due to his strong engagement in social networks, but for the other young business optimizers, this was not so visible.

Looking at the adult optimizers, DE3, DE4 and VE5 can be considered as progressive farmers from APL-plus, as they all owned large properties of land, on which they grew multiple crops. At the same time, they were not only increasing their livelihood portfolio in general, but also optimizing their business of training specifically, by increasing the number of cows or buffalo, or by including organic farming. In contrast, LE1 from BPL was not engaged in farming activities, yet she proactively developed her small Rangoli shop, since she had added a beauty parlour as a new business venture, which is why she is considered as a diversified optimizer. These optimizers had in common the fact that they too had connected the red, blue and yellow circles very neatly. They were not only targeted at increasing a single activity, but were able to think in a broader way, which was targeted at optimizing their diversified livelihood portfolio, for which they followed a business growth strategy. As for the drastic difference with respect to their social status, it must be noted that the APL-plus candidates had without a doubt a better starting position, as they did not face any financial constraints and easily gained access to credits, while LE1 used her limited financial resources for her business extension, without getting a loan from a bank, but she only borrowed some money.

The fact that only the adult entrepreneurs diversified their activities and the young entrepreneurs were mono-structured might be explained by the young age of the latter. This raises the concomitant question of whether business diversification would be a feasible option for them to further increase their income, and hence to enlarge their agency and better cope with limiting structures. At least in the case of DE9, business diversification was certainly an idea that he was entertaining, as he wanted to take further training to become more equipped to make things from milk, such as cheese or butter, which would in turn become a new business venture.

6.7.2 The diversifiers: the trial-and-error approach

As for the diversifiers, they have in common the fact that they were proactive when it came to their further activities, yet they were solely reactive when it came to developing their business of training. In this regard, they would not be very distinguishable from the preservers. They might not have been able to make their ends meet from their self-employed activity alone. In this sense, they proactively enlarged their agency, and thus they made use of their enabling structural conditions. Yet the additional activities were, except for DE8, disconnected from the self-employed activity in which they received the training. In addition to dairy farming, DE2 was running a general store, while DE6 was a home teacher with the goal of becoming a teacher at a government school. She seemed to follow a double strategy, as she was not willing to quit with dairy farming altogether. Due to the fact that she was very committed to dairy farming, and that both her activities were not necessarily incompatible, she was rather considered as a diversifier than someone following an education escape strategy. As for the case of SE3, the handbag-making business in which she conducted the entrepreneurship training was inactive, so she returned to her previous activity as a midwife, and, therefore, she is rather placed among the diversifiers than the conformists, who similarly did not manage to pursue her initially chosen self-employed activity. DE8 could have been considered as a preserver, as he showed passive behaviour patterns with respect to his support structure, even though he was very satisfied with the handholding with the training provider. Yet as a main entrepreneur, he also engaged in farming, which is why he was considered as a diversifier.

It can be concluded that the diversifiers from BPL who were engaged in disconnected activities followed more or less a trial-and-error approach. In this approach they only engaged in low-income-generating activities, which clearly served the goal of securing their survival. As for their broader future aspirations, they were very undetermined, not following a clear goal, e.g. by not wanting to invest in further education. In contrast, DE6 from APL had a clear educational aspiration. With respect to availing the support structure, they were all fairly disconnected in terms of engaging a farmers' club or SHG, or by obtaining help from financial support. Hence, they did not "thicken" their agency, although their performance in dairy farming was very different: DE2 had a high increase of income since training, DE6 a medium increase, and DE8 a low increase. Nevertheless, their proactive behaviour, when it came to further income-generating activities, widened their agency.

6.7.3 The career planners: relying on good education for a better job

The career planners have in common the fact that they clearly named a career aspiration apart from the self-employed activity, and that they were still young and had attained or were still pursuing higher educational degrees. In the cases of DE5 and DE10, they were destined to obtain a government job. For both of these entrepreneurs, this would mean quitting their self-employed activity. In the case of LE4, she was still very young and had just completed the twelfth grade at school. Even though she considered taking the training in beautification in order to open a beauty parlour, she was not placed among the business optimizers, as she was primarily focussed upon completing her school education. This could mean that she would still change her future aspirations, as she is in a transitional phase between school and work, and she had not yet developed any personal initiative with regard to her self-employed activity. In conclusion, the career planners can be distinguished from the preservers, since for them it was very likely that they would quit their self-employed activity due to their higher educational goals and professional aspirations. For this reason, the University degree failed to cohere for any of them with their self-employed activity, and as a result of this they were certainly not focussing upon a professionalization of their business activities. As for availing the support structure, they were all very disconnected when it comes to social engagement, availing financial support or handholding with the training provider, which means that they accepted limiting structural conditions with regard to their self-employed activity. Thus, they were not actively "thickening" their agency with regard to their self-employed activity. Because of this, they had a lower level of economic empowerment, but a high level of personal empowerment, which doubtless opened up new areas of actions for them to thrive in obtaining high-prestige career jobs. Yet it is questionable how realistic these high career aspirations really are for them to attain.

6.7.4 The preservers: maintaining the status quo

The preservers have in common the fact that they were all solely reactive with respect to coping with structural limitations. What stands out in this category is that they all came from a BPL-background, and with one exception attained low educational achievements. Furthermore, no dairy entrepreneur was assigned to that group, as these was the most diverse group of people. In the case of the preservers, they were mainly engaged in very low income-level activities with respect to possible earnings, e.g. as charkha workers or Rangoli shop-keepers.

The traditional handloom workers, namely HE1 and HE2, are both included in this group, as this was the one and only thing they had practised for their entire life. Yet, they seemed to opt to maintain the status quo instead of proactively developing the business forward. The Rangoli shop-keepers, who were LE2, LE3 and LE5, managed to keep the shop running on a minimum budget, but did not develop it further. In this respect, they were completely dependent on the training provider, who was overseeing the project and supplying them with goods. LE2 explicitly showed her reluctance to engage in social networks or to equip herself by taking further training. LE5, on the other hand, stands a little apart from this trend, as she completed a BA in political sciences, yet she was assigned to the preservers as she did not reveal any specific career aspiration or plan for how to use her degree. She also had no plan when it came to her self-employed activity or family tradition, which was handloom-working. Moreover, SE1 was interviewed while taking the stitching training. Considering the fact that she was already running a stitching business for many years, which certainly qualifies her as a preserver, she was at least ready to further equip herself by taking further training. However, she would not really match with the business optimizers, as, in her case, she had not yet received any opportunities to apply training inputs. She may develop personal initiative as a consequence of the training course, yet in considering her past occupations, she seemed to behave very reactively. She also considered applying for a job with the stitching certificate. Yet in this role she would not fit into the category of the business escapers, as she would remain in the same profession.

In conclusion, the preservers could be described as not having a clear vision for developing their business. Even though they all would probably agree that they wanted to increase their income, they were lacking any business growth strategy and rather conformed to their structural conditions, instead of "thickening" one's own agency and taking advantage of the existing support structure. Hence, they seemed to be lacking entrepreneurial competencies, and thus could – based on the distinction made by NIRD (2011) – be described as self-employed individuals rather than entrepreneurs. This is reflected in the strong degree of their disconnected pursuits, especially with respect to becoming engaged in a SHG or availing financial support.

6.7.5 The conformists: being undetermined and willing to do nothing and everything

The group of the conformists has been created for those entrepreneurs who took an entrepreneurship training, but were not able to build the business up due to various structural limitations. Even though both of the ladies assigned to this group came from BPL, they were very different personalities, FE1 being a young mother and holding a BA in Sociology, and SE2 being an older woman who had only completed fifth standard at school. FE1, who was already introduced in the case summaries, had not been assigned to the career planners, as she did not express any career aspirations. SE2, on the other hand, continued to work in stitching, but this reflected a completely different activity from making handbags, in which she had received the training. She was also following no specific goal, but was willing to take any training with respect to her education and skills, and was additionally willing to do any work which would improve her income. Both these candidates have in common the fact that they were strictly disconnected from the support structure. They had no handholding with the training provider, they did not engage socially, and SE2, at least, was not willing to apply for any loan. FE1 tried to apply for a loan, but was denied one, and she eventually ended up doing nothing, except being a mother. This is why these candidates are considered as conformist, as they only conformed to their structural conditions, instead of "thickening" their agency. Furthermore, they were both very undetermined with respect to their future aspirations. It seemed that they were willing to do either everything or nothing.

6.8 Analysing the relationships between types

As the final stage of the analysis, the relationships between the types, which have been analysed individually, will now be considered with regard to the specified research question, namely the question as to what livelihood strategies young entrepreneurs develop in relation to their occupational group, age, gender, social background and educational status, in order to cope with structural conditions, that enable or constrain them.

In considering the different occupational groups, it is noticeable that the dairy farmers are the most diversified group, since there only two out of ten were mono-structured. Five were proactively optimizing their dairy farming business, whereas the other five were either diversifying their livelihood portfolio with primarily disconnected activities, or they were following an education escape strategy. This is due to the low number of cows or buffalo, which is why this activity alone was not enough to make ends meet. Regarding the handloomers and mostly the Rangoli shop-keepers, who were mostly business preservers, it appears that being such a preserver was highly linked to their social status and occupational group. An important aspect of this is that of upholding traditions: handloom workers were very traditional, received the lowest educational achievements, and thus typically went on to become preservers. The dairy farmers were also traditionalists, but in a much more diverse way, since they were generally more open to change. For the surveyed charkha workers the activity was new, even though it was a very traditional activity. These women – and the same applies to the Rangoli shop-keepers – may be considered as traditionalists due to their social background, as they came from a village and a certain caste group. It needed the initiative of someone from outside, as was performed by Drishtee as part of its goal to develop sustainable communities, to introduce these women to such activities. This initiative was carried out in connection with the subject of women's empowerment. These women were given an opportunity to increase income, but without detaching them from their tradition, as these were home-based activities, and it allowed them to remain within their village environment.

As for the consideration of age, it is noteworthy that the young optimizers were only mono-structured, yet from these two examples no generalizations can be made. It is also not true that the others did not optimize. Except for DE10, all of the young entrepreneurs were planning to increase their dairy farming activity, yet only DE9 managed really to optimize it to a greater degree due to his strategic networks.

As for the question of gender, in every type-group both male and female representatives are found, which means that livelihood strategies did not overtly depend on gender. Gender is more of an issue of the occupational group, as in the beauty and wellness sector, only women were involved, as well as in spinning, while weaving happened to be a more male-dominated occupation. The gender breakdown of the dairy farming is particularly diverse, this usually being a family business in which both male and female members are engaged. With regard to the question of women's empowerment, many of the SHGs exist exclusively to empower women, yet none of the surveyed female entrepreneurs was engaged in such a group.

In considering the social background, great differences can be observed in all explored sectors. Within textiles, there were only entrepreneurs from BPL, i.e. in charkha, handloom and stitching. In Rangoli shop-keeping only BPL representatives were also engaged, with one exception. In contrast, the group of dairy farmers was very diverse, ranging from BPL to APL-plus. It is, moreover, noticeable that, with respect to the number of cows or monthly income, no significant differences between dairy farmers from BPL or APL were visible, but tremendous differences had occurred in the case of the APL-plus candidates, who increased their businesses on a larger scale. All other entrepreneurs kept their dairy farming business to a minimum scale of development. Among the preservers and conformists are only BPLers, while among the more educated escapers one finds only APLers, and among the proactive ones both seemed to be present equally.

In considering the educational status, it turned out that higher educated entrepreneurs who were pursing or had already attained a BA were counted in all identified type-groups. However, a clear distinction can be made between the optimizers, who had a higher educational degree, and the diversifiers or those following an education escape strategy. Whereas the higher educated optimizers did not plan to pursue their professional careers further, but rather anticipated that they would further develop their self-employed activity, the diversifiers had a clear goal apart from the self-employed activity. Yet they followed a double-strategy, wanting to pursue both objectives at the same time without planning to quit either one. In contrast, the escapers wanted to quit their self-employed activity and to attain a high-prestige job. Among the preservers, only LE5 had a BA degree, and she was not specific about her career aspirations, but simply followed routine instead of developing the business.

In conclusion, the entrepreneurs can generally be divided between preservers and optimizers, since the categories of diversification, education escape and conformity are only further specifications of the preservation strategy. Following this division, a bipolar picture emerges, which in turn helps to pose the key question: how should one best support the preservers to become optimizers? This question is relevant with respect to formulating recommendations for the training providers and further stakeholders. Yet the further sub-groups have been intentionally created, as this question must be considered for each type individually. For the escapers there is the question of business professionalization, as the university studies of these candidates seemed not to cohere with their self-employed activity: What university courses could be combined with self-employed activity? What additional training would be beneficial for them? On the other hand, for the diversifiers the question of appropriate training offers to further develop the business also arises, as well as the related question of how to develop an integrated growth strategy which is compatible with their different activities, as was the case with the diversified optimizers from APL-plus. For the optimizers in any of the surveyed professions, the basic question is how to scale the business activity. A good example is DE9, who thought ahead in terms of diversification and creating new ventures, e.g. by giving thought about how to do things from milk. These considerations provide a basis for the final part of this dissertation, which consists of the recommendations for the training providers as well as for societal and political stakeholders.

7. Conclusions

The main goal of the analysis carried out in the preceding chapters was to gain insights into the opportunities of empowering disadvantaged youth through entrepreneurship training. In considering these young entrepreneurs within the area of tension between agency and structure, together with their personal motivation and aspirations as they transition into adulthood, the analysis revealed that they developed different livelihood strategies, depending on their development of personal initiative and the degree of determination which they possessed in relation to their personal aspirations. Based on these findings, different types of entrepreneurs have been identified, which have specific implications for the training providers and further societal and political stakeholders that are engaged in the planning and realization of entrepreneurship training. In this chapter, firstly, lessons are drawn for the providers of entrepreneurship training concerning how to create opportunities of empowering the identified types of disadvantaged youth. Secondly, the focus will be shifted onto the role of further involved societal and political stakeholders, asking specifically how to change enabling and limiting structural conditions. Thirdly, the question will be elaborated how the present findings align with India's National Policy on Skill Development and Entrepreneurship. Finally, a look will be taken at possible future studies and directions of development. As the present analysis could only capture a fraction of the enormous issue of youth empowerment connected to entrepreneurship training, the question will be elaborated how development geography and its cognate disciplines could connect to the present analysis and enhance our present collective state of knowledge and policy.

7.1 Opportunities of empowering disadvantaged youth – drawing lessons for the training providers

The analysis has revealed a diverse picture with regard to youth empowerment. Only a few of the young entrepreneurs were empowered to thicken their agency and to change their limiting structural conditions, whereas most of them remained confined to their social structures. Based on these findings the relevant question is: How should one enhance entrepreneurship training as an instrument by which to foster the empowerment of disadvantaged youth? With regard to the different types that have been discovered in the course of the type-building analysis, lessons are drawn for the training providers regarding the major findings: How is one to foster personal initiative among the reactive entrepreneurs? How is one to scale home-based micro-enterprises? And what needs ought to be considered in light of the mismatch between personal aspirations and the self-employed activity? Finally, a plea is made for effective handholding that is sensitive to the needs of the poor.

7.1.1 How does one foster personal initiative?

With regard to dealing with the individuals' limiting or enabling structures, the question of how to foster personal initiative with regard to the self-employed activity must be considered in connection with the different types that have been identified on the basis of the individuals' different livelihood strategies. Obviously, this question is especially relevant for the reactive types of entrepreneurs, namely the preservers, the conformists, and the escapers. For the

© The Editor(s) (if applicable) and The Author(s), under exclusive license to Springer Fachmedien Wiesbaden GmbH, part of Springer Nature 2020
T. Aberle, *Entrepreneurship Training in Rural Parts of Bihar/ India*, Perspektiven der Humangeographie, https://doi.org/10.1007/978-3-658-30008-1_7

diversifiers, the question is relevant with respect to the development of their self-employed activity, in which they have to a great extent been passive, as they thrived to "thicken" their agency through further activities. Examples of this include DE2, who pursued an additional self-employed activity, or DE6, who was home-schooling. Thus, it all comes down to the key question: how does one give entrepreneurs, who have been passive and reactive with respect to the self-employed activity, the incentive to become proactive optimizers? If the proactive and reactive entrepreneurs are compared with each other, the following main ingredients can be identified that may foster personal initiative, and from which lessons can be derived for the training providers.

First of all, it seems that proactive entrepreneurs *identified themselves as competent actors* who can make a difference through action. This refers not only to being competent in carrying out the self-employed activity – which, apart from the inactive entrepreneurs, all of the surveyed entrepreneurs managed to do – but to change their limiting structural conditions through action. Thus, the individuals' identification as competent actors implies their knowledge about the limiting and enabling structural set-up. This is supported by Table 7 above (6.3), which showed a positive correlation between the development of personal initiative and one's awareness of structural limitations. Especially the candidates from APL-plus had, with their very diversified livelihood portfolio, a higher level of awareness regarding structural limitations, and thus certainly identified themselves as competent actors. Yet, knowledge alone is not sufficient to foster personal initiative, as many of the surveyed entrepreneurs had at least to a certain degree an understanding of the structural set-up, e.g. the availability of SHGs, but did not make use of it. Instead, it appeared that their motivation was blocked by their adherence to tradition and daily routines. Table 7 additionally revealed that proactive entrepreneurs were not only aware about the structural set-up, but also about their *scope of action*, which leads to the next point.

Secondly, proactive entrepreneurs must *know their scope of action* in connection with their self-employed activity. If they identify as competent actors and know how they could change the structural set-up through action, and thereby "thicken" their agency, they can foster personal initiative and act beyond their daily routines and tradition. This is how it appeared for the proactive entrepreneurs from both APL-plus and BPL. With their large possessions and diversified livelihood portfolios, the APL-plus candidates had without question the widest scope of action, which they proactively used for the optimization of their businesses. Likewise, the proactive entrepreneurs from BPL, such as DE9 and LE1, enhanced their agency by engaging in a leadership position in a farmers' club and by adding a beauty parlour to the Rangoli shop respectively. These examples show that, if the entrepreneurs only understand *how* they can change things through their actions, they become motivated to act accordingly, and thus develop personal initiative. This has implications for the training providers with respect to the training curricula and handholding with the entrepreneurs. It may be reasonable for entrepreneurship training, which primarily seems to focus on the imparting of technical know-how, to place greater emphasis on entrepreneurial aspects, which are practical and actionable, and which thus help to deal with constraining factors as well, depending on the occupational sector, the location and the social background of the entrepreneurs. Moreover, the training must include practical knowledge regarding how to scale the business (which will be discussed in

7.1.2) and how to avail the existing support structure, which leads in turn to the next point.

Thirdly, personal initiative can certainly be fostered through *community strength*, which becomes visible in community-based organizations, such as SHGs, famer's clubs, and cooperatives. In looking at the proactive entrepreneurs, it is noticeable that they were widely organized in networks, such as DE9, who was the paramount example as leader of a dairy farmers' club. However, the APL-plus candidates also understood the value of such networks with regard to supporting microentrepreneurs from BPL. DE4, from an APL-plus background, put it straightforwardly by saying that the poor should "become organized in a group and do everything in a group". Yet one finding from this survey was that, in the important aspects of gaining access to financial support and engaging in social networks, the training providers failed in most of the cases to provide any assistance or to use their connections to banks, farmers' clubs or SHGs (see 6.1.3.2). Hence, the following challenges occur for the training providers. On the one hand, they need to impart relevant information which is effective in action and which would specifically increase recursive knowledge regarding enabling or limiting structural conditions. This would include specific information about how to become engaged in SHGs, farmers' clubs and cooperatives and how to avail financial support, which would inform the individuals about where to apply for a loan, which would set out the criteria and procedures of different credit institutions. On the other hand, the training providers need to ensure effective handholding with the entrepreneurs, which would include the providing of active support in their becoming engaged in social networks and availing financial support. This will be discussed in further detail under 7.1.4. Yet many of these entrepreneurs seemed very dependent on Drishtee and missed the opportunity to develop their personal initiative, e.g. the Rangoli shop-keepers or the charkha workers. It appears that they perceived their agency only in fulfilling their duties and keeping the business running, and did not understand that they can become active themselves in areas where the duty of the training provider ends, e.g. in becoming socially engaged in collaborative networks. Therefore, the strong recommendation for the training providers remains to intentionally ensure effective assistance to young microentrepreneurs, with the goal of making them independent, and empowering proactive entrepreneurs to increase access to sustainable livelihoods. The sole lesson learned from the training is that it seems not to be enough to achieve this result, as the present analysis has revealed.

Fourthly, personal initiative can be fostered through *education and training*, which would obviously increase the individuals' level of competence, and with that their self-perception as competent actors, as well as consequently their scope of action. This became visible among the entrepreneurs from APL-plus, who all conducted various training at renowned training institutions. However, due to the fact that many of the surveyed entrepreneurs did take entrepreneurship training, but did not develop personal initiative, the question at hand remains: What was missing? How were they lacking in being unable to acquire entrepreneurial competencies? As has already been noted, another finding of the present analysis was that being trained does not automatically mean to be empowered. Many of the surveyed entrepreneurs took the training because they were recruited by a mobilizer of the training provider. Yet, even after the training, they were very dependent on the training provider and adhered to the daily routine, being confined with their existing structural conditions, and they continued to react in a solely reactive way. The question for these types of entrepreneurs is whether further training, which would further increase the level of competence in the individuals' respective business sector, would also foster more personal initiative. Taking into account the fact that a high proportion of the young entrepreneurs had a higher educational degree, which means they were young and capable to learn, it should be considered how they can be pushed to conduct further training. This will be further elaborated in consideration with scaling up the businesses (7.1.2).

In conclusion, it must be noted that the development of personal initiative usually means changing one's tradition and one's ways of doing things, which always requires taking risks, in which not all consequences can be seen in advance. These consequences can certainly not be accurately imparted for the individuals within the entrepreneurship training. Nevertheless, the effect of entrepreneurship training was undoubtedly that for some their confidence grew, as it was stated by BE1, which led to the development of personal initiative. Yet in achieving the goal of creating a force of empowered entrepreneurs, the identification of the individuals as competent actors, as well as developing a scheme of actionable knowledge regarding one's own scope of action, including how to scale the business in order to bring it to the next level, both seem to be a prerequisite to develop personal initiative. This will be further fostered through collaborative networks and by intentionally advancing the level of competence through further education and training.

7.1.2 How does one scale home-based microenterprises?

The question of how to scale home-based microenterprises is not only relevant for the optimizers, who are targeted at business growth, but to all the identified types of microentrepreneurs in various sectors, as long as they have already started to develop personal initiative. The present study revealed that the main motivation for engaging in microenterprises was to increase income. Yet most of the surveyed entrepreneurs simply stated that they wanted to increase their business, but were very imprecise about what that exactly meant for them and how they wanted to go about it. It seems that the majority had in fact no plan as to how to scale the business, which is understandable in view of their harsh financial, and perhaps also educational, constraints. Even though the training curricula (Appendices I-V), as well as the evaluation of both the entrepreneurs and the training providers, revealed that management aspects were also part of the training curriculum, it is still reasonable to ask whether aspects of how to scale the business were accurately captured. For instance, how does one develop a business growth plan? What is the best business model to use? Since these aspects surpass the scope of the present dissertation, at this point only some advice may be given for future related studies in order to further elaborate these aspects from a management perspective. Indeed, it seems appropriate in the present context to reflect on the question of what lessons could be drawn for microentrepreneurs from the APL-plus candidates, who managed to successfully scale up. Now, it is without any doubt that that, by being blessed with many possessions and finances, the APL-plus candidates had a much higher scope of action with respect to their business operations, and thus were able to obtain a higher area of influence and social recognition. Yet these candidates had some other common characteristics; they all had a diversified livelihood portfolio. In addition, they were engaged in the community, e.g. in the village panchayat or in political parties, and they were very committed at developing their vocational skills by attending various types of entrepreneurship training at renowned institutions. These aspects are now going to be considered with regard to their transferability for young microentrepreneurs from BPL. In this way, the different types of entrepreneurs will be more broadly kept in mind.

In looking at the exercise of building up a diversified business portfolio, the analysis revealed that – among the optimizers – only the adult entrepreneurs diversified their activities, whereas the young ones, by contrast, were mono-structured. This raises the question of whether business diversification would be a feasible option for them to further increase their income, and hence to enlarge their agency and better cope with limiting structures. The mono-structured portfolio might be explained by the young age of these entrepreneurs. Some were still unmarried and pursuing their education, which makes it difficult for them to put a lot of energy into building up a diversified livelihood portfolio. Yet, some of them were very committed and solely focussed at increasing their self-employed activity. Moreover, some even considered the opportunity of enhancing their level of competence through training in related areas, which would enable them to strategically diversify their business ventures, e.g. DE9, who considered to take training in how to make things from milk. Such a diversification strategy was successfully put into practice by LE1 from BPL, who took training in beautification and successfully added a beauty parlour to her Rangoli shop. These few examples show that for the microentrepreneurs it is, firstly, indeed appropriate to consider livelihood diversification as a strategy for business growth, e.g. by adding a new business venture, new ranges of products or new crops, where it makes sense to complement each other. Secondly, these stories drive home the point that such diversification must be accompanied by further training which is geared at acquiring both technical and entrepreneurial skills. This strategy may also be worth considering for diversifiers, such as DE2, who engaged in multiple disconnected businesses, i.e. in running a general store alongside his cattle raring. If both are operated at low scale and low margins, could other activities better fit into the livelihood portfolio?

Now, in looking at the aspect of social networking with regard to the microentrepreneurs, the exercise of engaging in community-based organizations, such as SHGs, farmers' clubs and corporations is certainly crucial for acquiring the necessary means in order to scale up. This has already been touched upon in consideration with the development of personal initiative. At this point, it seems appropriate to widen the focus somewhat by leaving the micro-level of individuals and by looking at microentrepreneurs as part of the "base-of-the-pyramid" market (henceforth BOP-market). New insights can be gained from the Indian economist C. K. Prahalad, who investigated the BOP-market specifically and recognized the great value of collaborative strength: "It is safe to say that the capacity to organize highly distributed and independent suppliers (for example, farmers) and contractors (for example, weavers) or potential consumers (for example, patients) and build a reliable logistics system is more critical than investment capital" (Prahalad 2014, 15). In his study Prahalad came to the conclusion that collaborative capacity is more valuable than investment capital. This broader perspective is helpful for the present analysis, since it widens the focus by looking not solely at the agency of individual entrepreneurs, but also at the effects on them societally as a whole. With this broader view in mind, it appears that Drishtee is without question one of the main actors involved in creating such logistics and networks at the BOP-market, e.g. by building a supply chain that provides a sustainable source of income to self-employed charkha and handloom workers, or by initiating the expansion and further creation of Rangoli shops. Now, in connection with the question of how to scale these businesses up, it must be recognized that the effects of entrepreneurship training may appear as not very significant for the individual, for instance if their monthly income increased by 50% from 4 000 to 6 000 Rupees.[98] However, the effects are still very considerable in their entire magnitude and reflect a valuable collaborative capacity, if this applies to thousands of microentrepreneurs. Not only that, many of them would not have any sustainable sources of income without the logistics system of the training provider. In looking at the great numbers of microentrepreneurs in rural Bihar, is it realistic at all to expect high business growth rates of the surveyed individual microentrepre-neurs? How much space for growth would the market allow until it reaches saturation? And yet, given the relatively low participation of the surveyed entrepreneurs in SHGs and the like, the collaborative capacity could certainly be further developed.

Finally, looking at the aspect of conducting further training, the question must be asked whether the training on offer is still relevant and up-to-date. In considering the development of current technology and the general trend towards globalization, it is reasonable to say that entrepreneurship training must catch up with these trends, which are for some of the evaluated sectors serious threats. For instance, in the case of the traditional handloom workers, it might be advisable to catch up with ongoing technologization and take further training that makes them qualified to use electrified handloom machines. This question is especially relevant for the career planners with higher educational degrees. Would it be feasible for them to conduct further training – just as the APL-plus candidates did – in the area of their self-employed activity at renowned training locations, which would match with their educational standard and help them to become more professional? However, the findings of the present analysis suggest that this must be a type of training which is addressed not only at further increasing the level of competence in the individuals' respective business sectors, but also at increasing their relevant business knowledge and entrepreneurial competencies. At any rate, if the APL-plus candidates from RSETI were given financial incentives in the form of credits for participating in the training, this should certainly also be made accessible to young and educated entrepreneurs from BPL, who hold a lot of potential.

7.1.3 Incoherency between the self-employed activity and personal aspirations: is there a need to adjust the range of training?

In the course of the present analysis, the youth sector has been intentionally considered in terms of their personal motivations and aspirations, which has provided especial insights regarding the livelihood strategies of young entrepreneurs. For some of the young entrepreneurs the analysis revealed a mismatch between their personal, professional, and educational aspirations on the one hand, and their self-employed activities, in which they conducted the entrepreneurship training, on the other hand. Three out of 14 candidates in the age cohort 18-35 have been identified as escapers, representing those who considered the self-employed

---

[98] Roughly from 50 to 75 Euros.

activity only as a Plan B. This trend was supported by a much higher share of the trainees.

The question now arises of how realistic the high career aspirations of these young adults are to actually reach, which in turn leads to the following considerations. Given these young adults' high educational levels, which ranged among the surveyed entrepreneurs up to Bachelor's degree level, and among the trainees up to pursuing a PhD, it stands out that the chosen university courses were highly incoherent with the self-employed activity. All of the surveyed entrepreneurs selected subjects in the Humanities, yet among the trainees at least three had intentionally chosen to study Economic Studies with the goal of advancing their self-employed activity. This leads to the question of what options such a business professionalization strategy holds to raise the self-employed activity on a higher level. This connects with the previous considerations which relate to how to scale the business, and will have to be considered for each occupational sector separately.

From the perspective of educational planning, it would be reasonable to scrutinize each of the target regions in order to identify the dominant agricultural and non-agricultural sectors in each case, with each of their needs for higher qualified staff, and to contrast it with the existing higher educational institutions and their available university courses. Now, that we know that a certain amount of young rural entrepreneurs are highly educated, but that they engage in very low-skilled self-employed activities with low margins, the question stands at hand whether some of them would have been better off to combine their studies with the self-employed activity, if they were given incentives, e.g. through specific scholarship programs. Yet the big incentive for government jobs and other high prestige jobs suggests that the agricultural sector, as well as traditional rural industries, are lacking in viable future prospects, and are therefore afflicted with a low image. Hence, many of them would likely not wish to change their career aspirations. Nevertheless, it is probably safe to say that some of them have to remain in the self-employed activity. For them, incentives should be given to allow them to pursue further entrepreneurship training, such as the APL-plus candidates did, especially in light of the high willingness that they exhibit to become further skilled, which was revealed during the present analysis.

The realization that many of the young entrepreneurs, who are highly educated but are stuck in microenterprises without knowing how to scale up, are willing to take further training and to settle in the long term for these occupations, leads to the question whether the range of training needs to be adjusted. All the visited training locations focussed on short-term entrepreneurship training, usually not exceeding 30 days, which imparted the very basics at a level where the only prerequisite was to be able to read and write. Such an educational setting may be appropriate for a high share of the low-educated rural youth cohort. Yet for the types of entrepreneurs who are represented by the escapers and optimizers, in which the majority, at least in the case of the dairy farmers, learnt the fundamentals already from scratch, the question can be asked whether such a training course really matches their level of competence and aspirations. It seems that they still longed for further training and advancement. Therefore, in light of Indian's qualification issue, skills gaps and the ambitious goal of skilling half a billion members of the youth sector by 2022, the onus is surely on India to come up with further advanced training for higher educated young adults. Such training could be imparted as a decentralized skilling mission, permeating rural parts of India, and lined up with PMKVY training courses. A first recommendation that can be derived for the training providers from this demand is to not only to look at numbers (how many are currently becoming enrolled in the training courses?), but to look more precisely at the young trainees' social and educational backgrounds, as well as existing levels of competence.

7.1.4 Towards an effective handholding, that is sensitive to the needs of the poor

From the previous elaborations the following conclusions can be derived for the training providers regarding effective handholding, that is sensitive to the needs of the poor. The challenge lies in keeping adequately in touch with the entrepreneurs, but at the same time ensuring and strengthening the entrepreneur's autonomy in not making him or her too dependent on the training provider, which would not encourage the development of personal initiative.

Certain general objections regarding handholding with the entrepreneurs after the training have been discovered, which have been expressed by different representatives of the training providers. Some of these objections relate to the issue of whose responsibility it is to ensure handholding. Statements of this kind include "The government is already given support" – which is at least true for PMKVY trainees, who receive a small financial grant afterwards – or "It is not part of my job description". Another objection relates to the fact that everything that a young entrepreneur needs to know in order to get started in his or her training is taught to him or her within the training period. As such the trainees can still approach the training provider to ask questions afterwards, but any active support will not be provided. A very valid concern raised by a stitching teacher is the following: "I cannot visit the students in all the places where they are working, it is a matter of the office." Some of these views clearly contradict the holistic community development approach at which all the surveyed training providers were targeted. If the goal is to truly *empower* disadvantaged youth through entrepreneurship training, some active kind of assistance must take place. Therefore, these objections will be addressed by the following suggestions regarding effective handholding.

First of all, all the staff engaged in the entrepreneurship training, whether as mobilizers, teachers or training managers, must understand that young entrepreneurs – especially those from a BPL background – are indeed in need of active support and assistance after the training. It would be wrong to assume that they will be able to manage because they were taught everything they need to know. This can be fixed as a key lesson from among the findings.

In light of the very diverse views and understandings of the concept of handholding, it seems important to have a clear definition of what 'handholding with the students' specifically means and whose responsibility it is, i.e. to establish concretely what is the role of the teachers, the mobilizers and the training managers.

Another issue is how to organize handholding in spite of the great diversity within the classroom. The range of trainees extends from BPL to APL-plus as well as from 'hardly being able to read and write' to 'pursuing a university degree'. A major constraint for all BPL candidates is that of availing financial support. Here, active support in helping them to avail the existing support network seems appropriate, e.g. for them to become engaged in SHGs, farmers' clubs or cooperatives. This may require the training providers to build networks with such community-based organizations, including banks. For the higher educated trainees, it might be advisable to focus upon further assistance on the further training options, that builds up from the inputs of the entrepreneurship training. This may require building up networks to further training institutions and to constantly keep up with new technological trends and developments.

From the previous findings it can be concluded that effective handholding must be targeted at fostering the development of personal initiative, which implies that the entrepreneurs must identify themselves as competent actors. This in itself reduces the training provider to a passive role, in which they cherish what the young entrepreneurs have already achieved and help them to understand the next step to take, in order to get the business started or advanced, encouraging them to do it. But it should be the entrepreneurs doing it, and not the training providers doing it for them.

7.2   The need for societal and political stakeholders to collaborate with the poor

In viewing young entrepreneurs within the area of tension between agency and structure, it was found that the entrepreneurs can proactively change their limiting structures as competent actors. Yet the area of influence in which they can make a difference through action is limited, as other stakeholders also play a crucial part in order to change the underlying structural conditions. Prahalad (2014, 26) stated, with regard to entrepreneurship, that "the opportunities at the BOP cannot be unlocked if large and small firms, governments, civil society organizations, development agencies, and the poor themselves do not work together with a shared agenda." Consequently, it can be concluded that opportunities of empowering disadvantaged youth through entrepreneurship training can only be ensured if all involved stakeholders collaborate together. Hence, the scope is now widened to include additional stakeholders, apart from the training providers. Thereby, emphasis is placed – in alignment with the final specified research question – on the recommendations that can be drawn for these stakeholders from the findings of the present analysis, with regard to the improvement of structural conditions for young entrepreneurs from socially deprived backgrounds. At first in what follows, various stakeholders are considered as part of the "ecosystem" (Prahalad 2014, 13), and brought in connection with the surveyed entrepreneurs. Next, recommendations will be derived for various stakeholders. Special attention will be given therein to the importance of granting credit for microentrepreneurs.

According to Prahalad, the environment in which microentrepreneurs operate can be described as an ecosystem, in which four types of actors can be identified: private enterprises, civil society organizations, development and aid agencies, and the BOP-entrepreneurs as well as BOP-consumers. His basic assumption is that these must collaborate with each other, so that "economic development" and "social transformation" can take place (Prahalad 2014, 26). In connection with the present analysis, these stakeholders have indeed been found to be very relevant for the surveyed entrepreneurs within the context of rural Bihar. A core finding was that many of the surveyed entrepreneurs were partly or completely disconnected from the available support structure, which could be conveniently paraphrased as an ecosystem. This will be illustrated with two contrasting examples from the case summaries (6.5). To cite a first example, the young lady from BPL, who took her training in inland fisheries (FE1), was completely detached from any support structure. She did not collaborate with private enterprises, such as banks who could have granted her a credit, or the owners of fish ponds. She was also not engaged in any community-based organization, such as a SHG or farmers' club, and she did not hold hands with the training provider or any other development agency. Consequently, she failed to start her business. In her case her ecosystem was completely lacking. On the contrary, DE9 was well connected with all parts of the ecosystem: as an individual, he was a BOP-entrepreneur, but as a leader of a dairy farmers' club he was highly engaged in community-based organizations. Furthermore, he was collaborating with private organizations, like banks or milk collectors, and he was also connected with the training provider. This leads to the question of what underlying factors caused the ecosystem to failure. So far, the focus has only been put on the entrepreneurs as competent actors. Apart from the training providers, further stakeholders have not been included within the survey for practical reasons grounded in limited capacities. Hence, no direct insights could have been gained from them. Yet, the example of FE1 has shown that they too play an active part within the ecosystem. For FE1 to succeed, it would have needed all engaged stakeholders to collaborate. The exact reasons why this collaborative network failed can certainly not be elucidated from this one particular case. Nevertheless, some general conclusions can be drawn for engaged stakeholders when it comes to becoming more sensitive to the needs of the poor, and ultimately to changing structural conditions in favour of the poor.

Microentrepreneurs must be recognized as competent actors as distinct from other engaged stakeholders, especially private enterprises of all kinds, e.g. large enterprises, small- and medium-sized enterprises, and banks. Prahalad found that the BOP-marked requires "a new respect for consumers as co-creators of solutions and not just passive recipients of a product or service" (Prahalad 2014, 15). This "respect for co-creators" must certainly also apply to BOP-entrepreneurs. BOP-entrepreneurs have the ability to create a tremendous degree of collaboration through community-based networks, and thus play a vital role in the ecosystem as a part of the solution of alleviating poverty. Yunus, the founder of the Grameen bank in Bangladesh, points out that "poverty is not caused by a person's unwillingness to work hard or lack of skill. As a matter of fact, a poor person works hard – harder than others, and he/she has more skill and time than he/she can use" (Yunus 1987, 3). This seems to accurately describe the stage of the poor, as it matches with this study's findings and observations themselves. It must only be added that acquiring the right skills is certainly crucial for the solution of the enormous issue of poverty in India and elsewhere. Thus, microentrepreneurs from a BPL background deserve to be valued and recognized by other societal and political stakeholders as "resilient and creative entrepreneurs" (Prahalad 2014, 25).

The co-creation with microentrepreneurs – to remain with this issue – would certainly foster personal initiative, not only among the microentrepreneurs but also among other stakeholders, as they start to recognize them as competent actors and credible business partners. This may result in the change of habits, where they have simply been ignored, or in the reconsideration of policies, for instance in the area of credit-lending. From the survey a few examples have been found where entrepreneurs from an APL-plus background seemed to understand this principle and proactively supported the poor. Take the example of DE4 from the Madhubani district, who helped small farmers by buying sugar cane from them in bulk purchases in order to deliver it to the sugar mill, which was 150 kilometres away and would be a hassle for the small-farmers to reach. Similarly, VE1, the entrepreneur who also came from the Madhubani district and who was engaged in vermicomposting, once started a farmers' club under NABARD and was also engaged in conveying training at Drishtee.

Microentrepreneurs must become eligible for credits. The major constraint of the surveyed young entrepreneurs concerned their lack of access to financial capital, which includes the grant of credits. This issue cannot simply be tackled as an economic issue alone. Yunus (1987) perceived it as a fundamental human rights issue, as it is connected with the rights to a decent standard of living.[99] This has severe implications for credit policies. Banks should not link the grant of credit to the condition of owning land, which would exclude a great share of the poor. At the same time, clear conditions must be in place, which ensure that loans are paid back in full and at their due time of repayment. The surveyed entrepreneurs have shown that they were generally very sincere in terms of financial matters. For instance, LE1, the Rangoli shop-keeper from the Madhubani district, returned the start-up money which was granted to her from Drishtee in full out of her own savings. Another woman said that she would not apply for a loan, as she was not sure whether she would be able to pay it back. This proves her sincerity. In addition, it points back to the previous consideration on how one ought in future to perceive the poor. Credit institutions must also recognize the poor as resilient and creative entrepreneurs.

7.3 Implications for bringing entrepreneurship training in alignment with India's skilling mission

In order to aid an unskilled workforce – of which less than 5% possessed any vocational skills[100] – and a large youth cohort and its resulting demographic dividend, which had great training demands, India adopted a National Skill Development Policy, with the aim of providing vocational education and training (henceforth VET) to half a billion people by the year 2022 (see 2.2.3.3). This policy was advanced in 2015 to become the National Policy for Skill Development and Entrepreneurship (NPSDE), emphasising that both job-oriented vocational education and entrepreneurship education were and are important for achieving the skilling task. The overall vision of the NPSDE is to "create an ecosystem of empowerment by skilling on a large scale at speed with high standards and to promote a culture of innovation-based entrepreneurship which can generate wealth and employment so as to ensure sustainable livelihoods for all citizens in the country" (MSDE 2015, 11). Furthermore, the NPSDE outlines the objective of fostering "innovation-driven and social entrepreneurship to address the needs of the population at the bottom of the pyramid" (MSDE 2015, 41). The entrepreneurship training that was part of the present analysis could be rightly captured under the term 'social entrepreneurship'.[101]

With regard to the findings of the present survey among young entrepreneurs, who belong to the mission of skilling half a billion people by 2022, implications will next be derived regarding the way that the recommendations made for the training providers and further engaged stakeholders (see 7.1 and 7.2) align with the core objectives outlined in the NPSDE,

---

[99] According to the Universal Declaration of Human Rights, Article 25 (1), adopted by the General Assembly of the United Nations on December 10, 1948 (see Yunus 1987, 1).

[100] Information taken from the National Policy on Skill Development and Entrepreneurship 2015 (MSDE 2015), based on formal skilling data for working age population from NSSO (68th Round) 2011-12.

[101] See Footnote 41 (2.2.3.6).

especially with regard to social entrepreneurship. In this inquiry, the following areas seem especially important. The first area concerns the proposed ecosystem of empowerment and the challenge of achieving the three types of skilling – skilling on a large scale, at speed, and at a high standard – at the same time. Does entrepreneurship training happen at the expense of quality? The second area concerns the challenge of ensuring innovative-based social entrepreneurship at the bottom of the pyramid. The third area concerns the demand of integrating entrepreneurship education into the formal educational system, which stems from the quest of raising the standard of training and tackling the tarnished image of vocational education.

7.3.1 Does entrepreneurship training happen at the expense of quality?

In light of the skilling task described above, concerns have been raised that the Indian VET-system is not fit to adequately respond to the current trends of the labour market (Beddie 2009; Dar 2008). The "ecosystem of empowerment" as described in the NPSDE – which is based on a collaboration of the public, private and voluntary sectors – follows the overall objective of skilling on a *large scale*, at *speed* and with *high standards* at the same time. These three traits, which certainly apply to entrepreneurship training as well, seem hardly to harmonize with each other, as the findings of the present analysis revealed. It appeared, for example, that the evaluated short-term entrepreneurship training courses were serving the purpose of – given the short duration of these courses – skilling at *speed*, and – given the low entrance barrier and the exhaustive recruitment strategy that is used to enjoin village after village – skilling on a *large scale*. Yet, a big question mark can be added when considering the envisaged outcomes of wealth, employment, and the insurance of access to sustainable livelihoods for all citizens. The evidence suggests that, for the majority of microentrepreneurs, it remains true that these self-employed activities hardly generate employment beyond one's own family. In addition, they rather serve to ensure their own subsistence than the creation of wealth. As a result, most of the entrepreneurs see little change or sense in gaining access to sustainable livelihoods, especially when it comes to property and land. Thus, the objective of skilling at a *high standard* seems to be lagging behind.

This finding resonates with the criticisms that the NPSDE attaches to providers of free training programs: "The various grant based, free training programmes available today, though necessary, have their own limitations especially on quality and employability. Students undergoing training for free attach little value to training whereas training providers focus on increasing their numbers rather than quality of training" (MSDE 2015, 8). Mathew (2017), a representative of Romesh Wadhwani's philanthropic initiative, which started a partnership with the central government's entrepreneurship development mission, expressed the criticisms with regard to short-term entrepreneurship training even more directly: "All these three-month skilling programmes are useless. You cannot take 12th graders with a power to earn '5 000 to 6 000'[102] a month, give them a month's training and equip them for a better task." He added that such a task requires at least twelve to 18 months of training.

The solution must be to raise the standard of training, which cannot be ensured without

---

[102] Indian Rupees, which is roughly between 60 and 75 Euros.

considering the great diversity in terms of social and educational background of the individuals involved, as well as personal aspirations of the trainees, that ranges from BPL to APL-plus and from 'being able to read and write' to University level. This must also go hand in hand with a collaboration of different stakeholders with the poor, as it has been outlined earlier. With regard to the great skilling needs that exist in India, it appears that a functioning ecosystem for entrepreneurship education, as it was outlined by Prahalad (2014, see 7.2), which caters to the needs of the BOP-market and which also ensures support for individual entrepreneurs through mentorship and networks, must be in place first, before the numbers can be increased. The goal of achieving inclusiveness cannot be achieved at the expense of quality.

7.3.2 Fostering innovation-driven entrepreneurship at the bottom of the pyramid

Another area of concern relates to the challenge of "fostering innovation-driven and social entrepreneurship to address the needs of the population at the bottom of the pyramid" (MSDE 2015, 41). As it was already outlined under 2.2.3.1, economists see innovation as a distinctive feature of entrepreneurship (Carland et al. 2007; Miller 1983; Schumpeter 1934). Private actors, such as Drishtee, have gained great expertise and enjoyed a healthy impact in their addressing of this challenge. Yet, based on the present survey, the concern can be raised that the characteristic of being innovation-driven certainly holds true for some of the training providers, but not necessarily for the microentrepreneurs, who were to a great deal confined within their structural limitations. It appeared that – with the exception of the dairy farmers – the collaborative strength lay only on the side of the training providers, which prevailed in terms of ideas, infrastructure, and in some cases also in the initial funding of the projects. However, this was prescribed to a high extent by the area of action of the individual entrepreneurs, e.g. in the cases of the Rangoli shop-keepers or charkha workers of Drishtee on the one hand, or in the case of the handbag makers of the Fateh Help Society on the other.

Hence, it seems important to elaborate the question of how personal initiative towards innovative-driven entrepreneurship among microentrepreneurs can be fostered. This may be achieved if the ecosystem pertains not only to linkages between various stakeholders and the training provider, but also to direct linkages between the stakeholders and the poor. This issue was, for example, evident at the RSETI-institution in the Madhubani district, which had good linkages with a renowned bank, yet in which poor entrepreneurs such as FE1 were excluded from gaining access to credit. However, in order to achieve this, the training provider must take on an active role as mediator between engaged stakeholders and microentrepreneurs from BPL.

7.3.3 Integrating entrepreneurship education into the formal education system

Another general issue at hand is that the Indian VET-system is confronted with a tarnished image, based on the prevailing concept that VET is a suitable option for the poor, offering a seemingly less favourable set of "second class" employment opportunities (Blaug 1973; Palanithurai 2016; Venkatram 2012). This results in a low degree of acceptance of the entrepreneurship training by the young population. For instance, at the school level the Indian Government opened vocational schools and envisaged diverting 25% of all enrolled children and youth into these schools, yet about a decade ago not even 1% were diverted into the vocational stream (Dar 2008). The fact that such a small number of the surveyed entrepreneurs aspired to self-employment as a viable career option suggests that entrepreneurship training for rural youth suffers from a tarnished image, due to its lack of future prospects in these activities, that cannot be maintained alongside educational and career aspirations, which are supported and promoted by the general education system.

In light of this situation, Venkatram (2012) proposes with regard to the VET-system in India to make vocational education an actual part of general education, in order to tackle its low public image and to recruit a higher share of the children and youth sector to become equipped with vocational skills. Yet, what would be the strategy for entrepreneurship training in remote and socially backward regions, such as rural Bihar, not only to reach a large share of the rural youth sector with short-term entrepreneurship training, but also to create a positive environment for self-employment, with profitable future prospects so that it can become worth aspiring for? The NPSDE formulates the objective of "integrating entrepreneurship education into the formal educational system" (MSDE 2015, 13). Their policy further emphasizes that a major key to the innovation ecosystem for entrepreneurship are entrepreneurship hubs across the country (E-hubs), among which are 3 000 colleges established to deliver entrepreneurship courses to enrolled students of various grades and courses. These are complemented by massively open online courses. More specifically, Universities and academic institutions are encouraged to launch a course on 'Social Entrepreneurship' as part of the strategy to integrate entrepreneurship education into the formal education system.

This idea would certainly address the issue of the present incongruity between University studies and the self-employed activities of young entrepreneurs. Given the fact that a high share of young entrepreneurs was pursuing or had completed a Bachelor's degree, many of them may take advantage of it. Yet, how can a subject named "social entrepreneurship" at University or college level be effective if it happens to be disconnected from the local communities? The high acceptance of entrepreneurship training candidates provided by Drishtee was due to Drishtee's connections forged with local communities, and a precise identification of occupations and skills requirements on the spot. Accordingly, such courses can only be effective if they are aligned with the relevant rural sectors, matching the realities in the villages, and if they offer opportunities of direct application into practice. Only then will they foster personal initiative. Yet, the challenge of creating networks with various stakeholders, and of integrating these networks into a viable ecosystem, also occurs outside the lecture hall or classroom. It therefore appears appropriate for the SHGs and farmers' clubs to become lined up with the respective Universities that are offering social entrepreneurship for its members, who take these courses. In this way, these group members can transfer their newly acquired knowledge to the less educated ones, which would build up collaborative strength.

A further aspect in this consideration of aligning entrepreneurship education with the formal education system is that many learn the self-employed activity already from scratch. Hence, it appears reasonable if entrepreneurship education does not start at college level, but already at the primary stage. This would be in line with holistic school models, which envisage including vocational education already in the school, such as the model of Bodhshala, which revolved around a production-integrated basic education programme in accordance with Gandhi's concept of basic education (Venkatesh 2014). Thus, the imparting of basic knowledge regarding relevant sectors, such as dairy farming, stitching or spinning, could start at the primary level, and managerial skills, with a specific focus on social entrepreneurship, could then be added at a secondary level. This may counteract the issue of the low appreciation of self-employed activities in agricultural and village-based sectors, raise the quality standards, and encourage entrepreneurship as a viable career option.

7.4 Concluding remarks

In considering entrepreneurship training as an instrument of empowering disadvantaged youth, it is clear that the demand for such training is at hand in light of the findings of the present analysis among youth from rural parts of Bihar, and in view of India's national skills development mission. Innovation-driven entrepreneurship must be fostered at the bottom of the pyramid. This requires entrepreneurship training targeted at fostering personal initiative. However, this cannot be realized at the expense of quality. The standard of entrepreneurship training must be raised, and microentrepreneurs must be included as competent actors within a viable ecosystem.

Yet, with regard to the increased educational levels of young rural entrepreneurs, combined with a mismatch between self-employed activities and University studies, the decision for such individuals to remain in traditional rural industries or agricultural activities stands in competition with the increasing educational options which entice them to thrive after migration and in career opportunities outside of their familiar settings. In order to consider this challenge in an appropriate manner, the disadvantaged youth sector has been approached from an action-oriented perspective within the area of tension between agency and structure. In this way, the qualitative research design allowed me to approach each entrepreneur on an individual level, taking into account his or her family background as well as his or her personal motivations, perceptions and aspirations.

However, this detailed and thoroughgoing evaluation required a narrowed focus that limited the scope of the survey to entrepreneurship training in rural parts of Bihar, with a manageable number of candidates from selected occupational sectors. Hence, only a fraction of the large issue at hand in relation to empowering disadvantaged youth through entrepreneurship training could be captured. This demands further related research, in both development geography as well as its cognate disciplines.

In view of the mass trends of globalization and rapid technological developments, which change the future workforce requirements for different occupational sectors and pose serious challenges with respect to the development of the agricultural sector, related research in neighbouring disciplines must address the issue of youth empowerment and entrepreneurship training. The following account tries to capture some noted areas of concern, which could not be covered by the present dissertation in detail, but which appear as important in gaining further insights regarding the issue at hand. In this light, it is firstly needless to say that these proposed connecting points claim no right of comprehensiveness. The economic sciences, e.g. development economics, agricultural economics or business management, could approach the topic of entrepreneurship training in rural parts of India by placing emphasis on the prospects for economic development on both the macro- and the micro-levels. At the macro-level, the BOP-market could be considered with regard to different occupational sectors, e.g. how to foster the development of agriculture and rural industries, but also how to develop the IT- and service-sectors in rural India along with the necessary infrastructure. On the micro-level, microentrepreneurs could be scrutinized regarding how they can scale up and achieve higher margins. Furthermore, the political sciences could place emphasis on issues of governance related to India's skilling task and how the public, private and voluntary sectors could work together in a more effective partnership. The social sciences, in particular sociology or psychology, could deepen insights into youth research, e.g. on youth transition into the world of work, the processes of socialization, and the personal perceptions, motivations and values of young microentrepreneurs from rural India, which affect their educational and professional choices. Not only that, the educational sciences could put education policies in a greater focus, e.g. by elaborating how the challenge of skilling on a large scale, at speed and at a high standard may be achieved. More specifically, they could examine how the standards of short-term entrepreneurship training could be raised, or from a didactic point of view, how the curricula could be brought into better alignment according to different needs represented among the very diverse group of trainees. Moreover, development geography and its cognate disciplines could elaborate on issues related to rural-urban disparities in relation to entrepreneurship education and the skilling task, emigration of rural youth from Bihar and its pulling factors, as well as issues of land distribution and its underlying social and political patterns. To qualify this last aspect, gaining access to land is clearly one of the most crucial issues for microentrepreneurs in rural Bihar.

This account could be further continued. As the present dissertation was placed within development geography, it appears appropriate to close this thesis by pointing to some implications for present-day development geography which concern how to address the issues mentioned above, especially in relation to youth empowerment, effectively. Present-day development geography has been described as action-oriented, multi-dimensional and transdisciplinary (see 2.1.1). These characteristics indeed appear as essential to the discipline's flourishing. Within an action-oriented approach, attention must be drawn to feasibility and practicability. Thus, research must be targeted at strengthening self-regulatory mechanisms, e.g. with respect to the microentrepreneurs, instead of reinforcing their dependency on government agencies or on development and aid agencies. This multi-dimensional approach puts the challenge of the researcher getting as close as possible to the object of research, which means being context-specific, understanding the perspectives of the concerned actors, and, at the same time, being sensitive to the greater societal context, e.g. in understanding the roles of multiple actors within the ecosystem for entrepreneurship education. Finally, it requires a transdiscipli-nary approach. Young people are in a dynamic and changeable stage of life, in which they may seem constantly in flux, regardless of where in the world they live. Hence, they cannot be approached as a static group or with static models, such as the sustainable livelihoods framework, but they must be seen at one dimension of a globalized world, which is growing together.

In conclusion, it should be noted that youth empowerment connected to entrepreneur-ship training is not only a complex field, offering various access points for transdisciplinary research, but is also highly relevant not just for Bihar or India but globally. It relates, as a matter of fact, to other pressing matters, such as youth unemployment or the refugee crisis, which demands that the root-causes of migration be addressed in the countries of the migrants' origin.

For these problems to be accomplished, it is inevitable that one should deal – as the present dissertation has done – with the future prospects of geographically and socially disadvantaged youth in developing countries. Hence, the issue of entrepreneurship education at the bottom of the pyramid deserves greater attention. Likewise, young entrepreneurs from lower sections of the society and from remote places – such as the selected target regions within rural Bihar – deserve it for their sakes that we "stop thinking of the poor as victims or as a burden and start recognising them as resilient and creative entrepreneurs" (Prahalad 2014, 25).

References

Alegre, I.; Kislenko, S.; Berbegal-Mirabent, J. (2017): Organized Chaos: Mapping the Definitions of Social Entrepreneurship. In: Journal of Social Entrepreneurship 8 (2): 248-264.

Aranguri, C.; Davidson, B.; Ramirez, R. (2006): Patterns of Communication through Interpreters: A Detailed Sociolinguistic Analysis. In: Journal of General Internal Medicine 21: 623-629.

ASCI – Agriculture Skill Council of India (n.a.): LMIS Report on Dairy Sector. http://asci-india.com/pdf/LMIS-on-Dairy.pdf (October 17, 2017)

Babu, B. R. (2011): Education and Ideology of Gandhi and Ivan Illich. New Delhi: Neelkamal Publications.

Bauer, U. (2012): Sozialisation und Ungleichheit. Eine Hinführung. 2nd Edition. Wiesbaden: VS Verlag für Sozialwissenschaften.

Beck, U. (1986) Risikogesellschaft: Auf dem Weg in eine andere Moderne. 1st Edition. Frankfurt am Main: Suhrkamp.

Becker, H. S.; Geer, B. (1979): Teilnehmende Beobachtung: Die Analyse qualitativer Forschungsergebnisse. In: Hopf, C.; Weingarten, E. (eds.): Qualitative Sozialforschung. 1st Edition. Stuttgart: Klett-Cotta: 139-168.

Beddie, F. M. (2009): Australia and India: Facing the Twenty-first Century Skills Challenge. New Delhi: NCVER (Paper presented at the annual Global Skills Summit: Emerging Skills – India 2020).

Behera, D; Chaudhary, A.K.; Vutukuru, V.K.; Gupta, A.; Machiraju, S.; Shah, P. (2013): Enhancing Agricultural Livelihoods through Community Institutions in Bihar, India. Bihar Livelihoods Promotion Society. Washington D.C: The World Bank. http://documents.worldbank.org/curated/en/2013/01/17517688/enhancing-agricultural-livelihoods-through-community-institutions-bihar-india (September 5, 2017)

Behrendt, R. F. (1969): Soziale Strategie für Entwicklungsländer: Entwurf einer Entwicklungssoziologie. 2nd Edition. Frankfart a. Main: Fischer.

Bell, S.; Payne, R. (2009): Young People as Agents in Development Processes: Reconsidering Perspectives for Development Geography. In: Third World Quarterly 30 (5): 1027-1044.

Bendit, R. (2008): Looking to the Future: Some New Perspectives for Youth Research and Knowledge based Youth Policy. In: Bendit, R.; Hahn-Bleibtreu, M. (eds.): Youth Transitions. Processes of Social Inclusion and Patterns of Vulnerability in a Globalised World. Opladen. Farmington Hills: Budrich: 359-372.

Berger, P. L.; Luckmann, T. (1967): The Social Construction of Reality. 1st Edition. London: Lane The Penguin Press.

© The Editor(s) (if applicable) and The Author(s), under exclusive
license to Springer Fachmedien Wiesbaden GmbH, part of Springer Nature 2020
T. Aberle, *Entrepreneurship Training in Rural Parts of Bihar/ India*, Perspektiven
der Humangeographie, https://doi.org/10.1007/978-3-658-30008-1

Berman, R. C.; Tyyskä, V. (2011): A Critical Reflection on the Use of Translators/Interpreters in a Qualitative Cross-Language Research Project. In: International Journal of Qualitative Methods 10 (1): 178-190.

Bhagat, R. B. (2016): Changing Pattern of Internal Migration in India. In: Guilmoto, C. Z.; Jones, G. W. (eds.): Contemporary Demographic Transformations in China, India and Indonesia. Heidelberg. New York. Dordrecht. London: Springer: 239-254.

Blaug, M. (1973): Education and the Employment Problem in Developing Countries. Geneva: ILO.

Blenck, J. (1979): Geographische Entwicklungsforschung. In: Hottes, K.-H. (ed.): Geographische Beiträge zur Entwicklungs-Länderforschung. DGFK-Hefte 12: 11-20.

Bohle, H.-G. (2011): Vom Raum zum Menschen. Geographische Entwicklungsforschung als Handlungswissenschaft. In: Gebhardt, H.; Glaser, R.; Radtke, U.; Reuber, P. (eds.): Geographie, Physische und Humangeographie. 2nd Edition. Heidelberg: Spektrum Akademischer Verlag: 746-763.

Bose, A. (2005): No Push-Button Solutions. In: The Tribune, September 24, 2005. Chandigarh. http://www.tribuneindia.com/2005/specials/tribune_125/main19.htm (October 5, 2017)

Bourdieu, P. (1979): Entwurf einer Theorie der Praxis auf der ethnologischen Grundlage der kabylischen Gesellschaft. 1st Edition. Frankfurt am Main: Suhrkamp.

Bronfenbrenner, U. (1981): Die Ökologie der menschlichen Entwicklung. 1st Edition. Stuttgart: Klett-Cotta.

Bronger, D.; Trettin, L. (eds.) (2008): Megastädte – Global Cities Heute: Das Zeitalter Asiens? Münster: LIT.

Bruyat, C. A.; Julien, P.A. (2001): Defining the Field of Research in Entrepreneurship. In: Journal of Business Venturing 16: 165-180.

Burt, R. S. (2000): The Network Structure of Social Capital. In: Research in Organizational Behaviour 22: 345-423. https://www.bebr.ufl.edu/sites/default/files/The%20Network%20Structure%20of%20Social%20Capital.pdf (August 3rd, 2018)

Cappai, G. (2003): Grundlagentheoretische und methodologische Bemerkungen zum Interpretieren und Übersetzen als interkulturelle Operationen. Für einen möglichen Dialog zwischen analytischer Philosophie und Sozialwissenschaften. In: Ziegerle, A.; Cappai, G. (eds.): Sozialwissenschaftliches Übersetzen als interkulturelle Hermeneutik. Berlin: Duncker & Humblot: 107-131.

Carland, J. W.; Hoy, F.; Boulton, W. R.; Carland, J. A. C. (2007): Differentiating Entrepreneurs from Small Business Owners: A Conceptualization. In: Cuervo, Á.; Ribeiro, D.; Roig, S. (eds.): Entrepreneurship: Concepts, Theory and Perspective. Berlin. Heidelberg: Springer: 73-81.

Carney, D; Drinkwater, M.; Rusinow, T.; Neefjes, K.; Wanmali, S.; Singh, N. (1999): Livelihood Approaches Compared: A Brief Comparison of the Livelihoods Approaches of the UK Department for International Development (DFID), CARE, Oxfam and the United Nations Development Programme (UNDP). London: DFID.

Castells, Manuel (2010): The Rise of the Network Society. 2nd Edition. Malden: Wiley-Blackwell.

Chambers, R. (1989): Editorial Introduction: Vulnerability, Coping and Policy. In: IDS Bulletin 20 (2): 1-7.

Chambers, R.; Conway, G. R. (1991): Sustainable Rural Livelihoods: Practical Concepts for the 21st Century. IDS Discussion Paper 296, December 1991. Brighton: Institute of Development Studies. www.ids.ac.uk/files/Dp296.pdf (June 20, 2017)

Chenoy, D. (2013): Skill Development in India: A Transformation in the Making. In: IDFC Foundation (ed.): India Infrastructure Report 2012. New Delhi: IDFC Foundation: 199-207.

Census of India – Office of the Registrar General and Census Commissioner, Government of India (2011): Age Data: Single Year Age Data - C13 Table (India/States/UTs) http://www.censusindia.gov.in/2011census/Age_level_data/Age_level_data.html (August 2nd, 2018)

CWMG – Collected Works of Mahatma Gandhi (1888-1948). Sevagram: Sevagram Ashram. http://www.gandhiashramsevagram.org/gandhi-literature/collected-works-of-mahatma-gandhi-volume-1-to-98.php (October 29, 2017) Volumes, which have been referred to:
CWMG , Vol. 30 (December 27, 1924 - March 21, 1925)
CWMG , Vol. 66 (December 16, 1934 - April 24, 1935)
CWMG , Vol. 72 (July 6, 1937 - February 20, 1938)
CWMG , Vol. 91 (May 20, 1946 - August 8, 1946)

Dar, A. (2008): Skill Development in India: The Vocational Education and Training System. South Asia Human Development Sector Series (22). Washington, D.C.: World Bank.

Dasgupta, P. K.; Haralkar, S. N.; Singh, S. B. (2010): Continued Relevance of Basic Education. In: Prabhath, S. V. (ed.): Perspectives on Nai Talim. New Delhi: Serials Publications: 67-73.

Dayal, P. (2013): The Agricultural Geography of Bihar. New Delhi: Rajesh Publications.

Deffner, V.; Haferburg, C. (2014): Bourdieus Theorie der Praxis als alternative Perspektive für die "geographische Entwicklungsforschung". In: Geographica Helvetica 69 (1): 7-18.

Deshpande, S. (2014): The Problem of Caste. New Delhi: Orient Blackswan.

DeSouza, P. R.; Kumar, S.; Shastri, S. (2009): Indian Youth in a Transforming World. Attitudes and Perceptions. New Delhi: Sage Publications.

DFID – Department for International Development UK (1999): Sustainable Livelihood Guidance Sheet. London: DFID. Image taken from: http://practicalaction.org/images/sl-framework-colour.gif (April 15, 2016)

Dörfler, T.; Graefe, O.; Müller-Mahn, D. (2003): Habitus und Feld: Anregungen für eine Neuorientierung der geographischen Entwicklungsforschung auf der Grundlage von Bourdieus Theorie der Praxis. In: Geographica Helvetica 58 (1): 11-23.

Drishtee (n.a.): Homepage of Drishtee. http://www.drishtee.com (June 1, 2018)

Encyclopedia Britannica (2016): Ahīr: Hindi Subcaste. Homepage of the Encyclopaedia Britannica, Inc. https://www.britannica.com/topic/Ahir (August 2, 2018)

Erikson, E. H. (1968): Identity, Youth and Crisis. New York: Norton.

Erikson, E. H. (1974): Jugend und Krise: Die Psychodynamik im sozialen Wandel. 2nd Edition. Stuttgart: Klett.

Escobar, A. (1995): Encountering Development: The Making and Unmaking of the Third World. Princeton: Princeton University Press.

Feroze, S. M.; Chauhan, A. K. (2010): Performance of Dairy Self Help Groups (SHGs) in India: Principal Component Analysis (PCA) Approach. In: Indian Journal of Agricultural Economics 65 (2): 308-320.

Flick, U. (1987): Methodenangemessene Gütekriterien in der qualitativ-interpretativen Forschung. In: Bergold, J. B.; Flick, U. (eds.): Ein-Sichten. Tübingen: DGVT: 247-262.

Flick, U. (1989): Vertrauen, Verwalten, Einweisen. Subjektive Vertrauenstheorien in sozialpsychiatrischer Beratung. Opladen: Deutscher Universitätsverlag.

Flick, U. (1999): Qualitative Forschung: Theorie, Methoden, Anwendung in Psychologie und Sozialwissenschaften. Reinbek bei Hamburg: Rowohlt.

Frese, M. (2000): Executive Summary, Conclusions, and Policy Implications. In: Frese, M. (ed.): Success and Failure of Microbusiness Owners in Africa: A Psychological Approach. Westport: Greenwood Publishers: 161-190.

Frese, M.; Fay, D. (2000): Entwicklung von Eigeninitiative: Neue Herausforderung für Mitarbeiter und Führungskräfte. In: Welge, M. K.; Häring, K.; Voss, A. (eds.): Management Development: Praxis, Trends und Perspektiven. Stuttgart: Schäffer-Poeschel: 63-78.

Frese, M. (2009): Toward a Psychology of Entrepreneurship: An Action Theory Perspective. In: Foundations and Trends in Entrepreneurship 5 (6): 435-494.

Friedrichs, J. (1985): Methoden empirischer Sozialforschung. 13th Edition. Opladen: Westdeutscher Verlag.

Gandhi, M. K. (1951): Basic Education. Ahmedabad: Navajivan Publishing House.

Ganesh, A. (2010): Self Employment Training: An Approach to Women Empowerment. In: Sing, A. K., Singh, S. P., Sutaria, D.S. (eds.): Gender Budgeting and Woman Empowerment in India. New Delhi: Serials Publications: 154-168.

Garba, P. K. (1999): An Endogenous Empowerment Strategy: A Case Study of Nigerian Woman. In: Development in Practice 9 (1-2): 130-141.

Gengaiah, U. (2016): NGO Initiatives: Non-Governmental Organisation Initiatives. In: Pilz, M (ed.): India: Preparation for the World of Work – Education System and School to Work Transition. Wiesbaden: Springer VS: 211-229.

Giddens, A. (1979): Central Problems in Social Theory. Action, Structure and Contradiction in Social Analysis. Berkeley: University of California Press.

Giddens, A. (1984): The Constitution of Society: Outline of the Theory of Structuration. Berkeley. Los Angeles: University of California Press.

Giddens, A. (1990): The Consequences of Modernity. 1st Edition. Cambridge: Polity Press.

Girtler, R. (1984): Methoden der qualitativen Sozialforschung. Wien: Böhlau.

GoB – Government of Bihar, Financial Department (n.a.): Economic Survey 2016-17. Patna. http://finance.bih.nic.in/Documents/Reports/Economic-Survey-2017-EN.pdf (November 2, 2017)

GoI – Government of India (2003): National Youth Policy 2003. http://www.youthpolicy.org/national/India_2003_National_Youth_Policy.pdf (October 15, 2017)

GoI – Government of India (2008): Bihar`s Agricultural Development: Opportunities and Challenges. A Report of the Special Task Force on Bihar. New Delhi. http:// http://planningcommission.gov.in/aboutus/taskforce/tsk_adoc.pdf (July 22, 2018)

Grundmann, M. (1994): Das Scheitern der sozialstrukturellen Sozialisationsforschung oder frühzeitiger Abbruch einer fruchtbaren Diskussion? In: Zeitschrift für Sozialisations-forschung und Erziehungssoziologie 14 (2): 163-186.

Grundmann, M.; Hoffmeister, D. (2007): Die Verwobenheit von Sozialisation und Selektion. In: Zeitschrift für Soziologie der Erziehung und Sozialisation 27 (2): 2-20.

Havighurst, R. J. (1953): Human Development and Education. 1st Edition. New York: Longmans, Green.

Hurrelmann, K. (1986): Einführung in die Sozialisationstheorie. Über den Zusammenhang von Sozialstruktur und Persönlichkeit. 1st Edition. Weinheim: Beltz.

Hurrelmann, K. (1988): Social Structure and Personality Development. The Individual as a Productive Processor of Reality. Cambridge: Cambridge University Press.

Hurrelmann, K. (1995): Lebensphase Jugend. Eine Einführung in die sozialwissenschaftliche Jugendforschung. 4th. Edition. Weinheim. München: Juventa-Verlag.

Hurrelmann et al. (2008): Hurrelmann, K.; Matthias, G.; Sabine, W. (eds.) (2008): Handbuch Sozialisationsforschung. 7th Edition. Weinheim: Beltz.

Hurrelmann, K. (2012): Sozialisation: Das Modell der produktiven Realitätsverarbeitung. 10th Edition. Weinheim. Basel: Beltz.

Hurrelmann, K.; Quenzel, G. (2016): Lebensphase Jugend. Eine Einführung in die sozialwissenschaftliche Jugendforschung. 13th Edition. Weinheim. Basel: Beltz Juventa.

Hussain, M. D., Bhuiyan, A. B., Bakar, R. (2014): Entrepreneurship Development and Poverty Alleviation. An Empirical Review. In: Journal of Asian Pacific Research 4 (10): 558-573.

IFAD – International Fund for Agricultural Development (2005): Empowerment of Poor. Enabling the Rural Poor to Overcome Poverty. https://www.ifad.org/event/past/tags/2184829 (April 15, 2016)

ILFS – Infrastructure Leasing and Financial Services Limited (2007): A Report on Diagnostic Survey and Business Plan for Handloom Sector in Bihar. Submitted to Department of Industries, Government of Bihar. http://industries.bih.nic.in/acts/ad-02-24-04-2007.pdf (October 7, 2017)

Islam, M. N.; Imam, A. (2011): SHG Model of Indian Microfinance: Endeavors and Challenges. In: Khan, N. U.; Sigamani, P. (eds.): Anatomy of Public Reforms and Development. New Delhi: MacMillan Publishers: 314-322.

Jan, M. (2009): Agricultural Factors Associated with Women Empowerment. In: Roy, U.N.; Saini, J.S. (eds.): People's Empowerment and Sustainable Rural Development. A Technological Approach. Jaipur. New Delhi: 80-89.

Jick, T. D. (1983): Mixing Qualitative and Quantitative Methods: Triangulation in Action. In: Maanen, J. V. (ed.): Qualitative Methodology. Beverly Hills: Sage Publications: 135-148.

Kallmeyer, W.; Schütze, F. (1976): Konversationsanalyse. In: Studium Linguistik 1: 1-28.

Kapborg, I.; Berterö, C. (2002): Using an Interpreter in Qualitative Interviews: Does it Threaten Validity? In: Nursing Inquiry 9 (1): 52-56.

Keniston, K. (1968): Young Radicals. New York: Harcourt, Brace and World.

Klocker, N. (2007): An Example of "Thin" Agency: Child Domestic Workers in Tanzania. In: Panelli, R.; Punch, S.; Robson, E. (eds.): Global Perspectives on Rural Childhood and Youth: Young Rural Lives. New York: Routledge: 83-94.

Klüver, J. (1979): Kommunikative Validierung – einige vorbereitete Bemerkungen zum Projekt "Lebensweltanalyse von Fernstudenten". In: Heinze, T. (ed.): Theoretische und methodologische Überlegungen zum Typus hermeneutisch-lebensgeschichtlicher Forschung. Hagen: Fernuniversität: 69-84.

Knutsson, P. (2006): The Sustainable Livelihoods Approach: A Framework for Knowledge Integration Assessment. In: Human Ecology Review 13 (1): 90-99.

KPMG (n.a.): Skilling India: A Look Back at the Progress, Challenges and the Way Forward. http://ficci.in/spdocument/20405/FICCI-KPMG-Global-Skills-report.pdf (October 12, 2017)

KPMG, Volume 1 (n.a.): Human Resources and Skill Requirement in the Agriculture Sector (2013-17, 2017-22). http://www.nsdcindia.org/sites/default/files/files/Agriculture.pdf (October 19, 2017)

KPMG, Volume 4 (n.a.): Human Resources and Skill Requirement in the Beauty and Wellness Sector (2013-17, 2017-22). https://www.nsdcindia.org/sites/default/files/files/Agriculture.pdf (October 19, 2017)

Krantz, L. (2001): The Sustainable Livelihood Approach to Poverty Reduction. An Introduction. Stockholm: Swedish International Development Cooperation Agency. https://www.sida.se/contentassets/bd474c210163447c9a7963d77c64148a/the-sustainable-livelihood-approach-to-poverty-reduction_2656.pdf (June 6, 2017)

Krüger, F. (2003): Handlungsorientierte Entwicklungsforschung: Trends, Perspektiven, Defizite. In: Petermanns Geographische Mitteilungen 147 (1): 6-15.

Kruse, J.; Bethmann, S.; Niermann, D.; Schmieder, C. (eds.) (2012): Qualitative Interviewforschung in und mit fremden Sprachen: Eine Einführung in Theorie und Praxis. Weinheim. Basel: Beltz Juventa.

Kuckartz, U. (2014): Qualitative Text Analysis. A Guide to Methods, Practice and Using Software. London: Sage Publications.

Kuckartz, U. (2016): Qualitative Inhaltsanalyse. Methoden, Praxis, Computerunterstützung. 3rd Edition. Weinheim. Basel: Beltz Juventa.

Kulke, E.; Staffeld, R. (2009): Informal Production Systems: Plastic Recycling andProcessing in Dhaka. In: Die Erde 140 (1): 25-43.

Kumar, N.; Bhagat, R. B. (2016): Social Network and Mass Exodus of Youth from Rural Bihar. In: Roy, P. M., Ghosh, A. (eds.): Rural Development – Contemporary Issues and Practices. Kolkata: Rohini Nandan: 228-247.

Kurz, H. D. (1994): Ricardo and Lowe on Machinery. In: Wood, J. C. (ed.): David Ricardo – Critical Assessment. Second Series (6). London: Routledge: 88-110.

Lang-Wojtasik, G. (1997): Theorie und Praxis von Gandhis Basic Education als Perspektive zukunftsfähiger Bildungskonzeptionen. In: Noormann, H.; Lang-Wojtasik, G. (ed.): Die Eine Welt der vielen Wirklichkeiten. Pädagogische Orientierungen. Festschrift für Asit Datta. Frankfurt am Main: IKO: 211-230.

Lang-Wojtasik, G. (2002): Gandhis Nai Talim im Kontext von Education for All. In: Datta, A.; Lang-Wojtasik, G. (ed.): Bildung zur Eigenständigkeit: Vergessene reformpädagogische Ansätze aus vier Kontinenten. Frankfurt am Main. London: IKO: 185-200.

Larkin, P. J.; de Casterlé, B. D.; Schotsmans, P. (2007): Multilingual Translation Issues in Qualitative Research: Reflections on a Metaphorical Process. In: Qualitative Health Research 17 (4): 468-476.

Lewis, M. P.; Simons, G. F.; Fennig, C. D. (eds.) (2015): Ethnologue. Languages of the World. 18th edition. Dallas: SIL International. https://www.ethnologue.com/18/country/IN/languages/index-2.html (April 28, 2018).

Low, M. B.; MacMillan, I. C. (2007): Entrepreneurship: Past Research and Future Challenges. In: Cuervo, Á.; Ribeiro, D.; Roig, S. (eds.): Entrepreneurship: Concepts, Theory and Perspective. Berlin. Heidelberg: Springer: 131-154.

Lumpkin, G. T.; Dess, G. G. (1996): Clarifying the Entrepreneurial Orientation Construct and Linking it to Performance. In: Academy of Management Revue 21 (1): 135-172.

Mandelman, F. S.; Montes-Rojas, G. V. (2009): Is Self-Employment and Micro-Entrepreneurship a Desired Outcome? In: World Development 37 (12): 1914-1925.

Mangold, W. (1960): Gegenstand und Methode des Gruppendiskussionsverfahrens: Aus der Arbeit des Instituts für Sozialforschung. Frankfurt am Main: Europäische Verlagsanstalt.

Mathew, J. C. (2017): "All these Three-Month Skilling Programmes are Useless": On February 23, the Ministry of Skill Development and Entrepreneurship announced a Partnership with US-based Billionaire Romesh Wadhwani's Philanthropic Initiative, Wadhwani Operating Foundation, for the Pradhan Mantri Yuva Udyamita Vikas Abhiyaan (YUVA), the Central Government's Entrepreneurship Development Mission. In: Business Today, April 9, 2017. New Delhi: Living Media India, Limited. https://www.businesstoday.in/magazine/columns/all-these-three-month-skilling-programmes-are-useless/story/24 8323.html (July 3, 2018)

Mattissek, A.; Pfaffenbach, C.; Reuber, P. (2013): Methoden der empirischen Humangeographie. Braunschweig: Westermann.

Mayer, H. O. (2013): Interview und schriftliche Befragung: Grundlagen und Methoden der empirischen Sozialforschung. München: Oldenbourg.

Mayer, M. (2002a): Jugend, Verwundbarkeit und soziale Diskriminierung. Lebenschancen und Konfliktpotentiale ländlicher Jugendlicher in Sri Lanka. In: Geographica Helvetica 57 (1): 19-33.

Mayer, M. (2002b): Jugendkonflikte und Entwicklungsplanung. Eine sozialgeographische Analyse der Lebenschancen Jugendlicher und der Potentiale dezentraler Planung in Sri Lanka. Saarbrücken: Verlag für Entwicklungspolitik.

Mayring, P. (2015): Qualitative Inhaltsanalyse. Grundlagen und Techniken. 12th Edition. Weinheim: Beltz.

Mayring, P. (2016): Einführung in die qualitative Sozialforschung. 6th Edition. Weinheim: Beltz.

Mehrotra, S. (2016): Realising the Demographic Dividend: Policies to Achieve Inclusive Growth in India. New Delhi: Cambridge University Press.

Merriam-Webster Dictionary (2017): Definition of Agent. https://www.merriam-webster.com/dictionary/agent (October 31, 2017)

Mies, M. (1993): The Myth of Catching-up Development. In: Mies, M; Vandana, S. (eds.): Ecofeminism. Halifax, Nova Scotia: Fernwood Publications: 55-69.

Miles, M. B.; Huberman, A. M.; Saldana, J. (2014): Qualitative Data Analysis. A Methods Sourcebook. 3rd Edition. Los Angeles. London. New Delhi. Singapore. Washington D.C: Sage.

Miller, D. (1983): The Correlates of Entrepreneurship in Three Types of Firms. In: Management Science 29 (7): 770-791.

Mitra, A.; Verick S. (2013): Youth Employment and Unemployment: An Indian Perspective. Asia-Pacific Working Paper Series, DWT for South Asia and Country Office for India. Geneva: ILO. http://www.ilo.org/wcmsp5/groups/public/---asia/---ro-bangkok/---sro-new_delhi/documents/publication/wcms_211552.pdf (October 27, 2017)

Mitra, S.; Nagarajan, R. (2005): Making Use of the Window of Demographic Opportunity: An Economic Perspective. In: Economical and Political Weekly, December 10, 2005: Sameeksha Trust.

MoF – Ministry of Finances, Government of India (2007): Report of the Expert Group on Agricultural Indebtedness. New Delhi. http://www.igidr.ac.in/pdf/publication/PP-059.pdf (October 22, 2017)

Mohanty, B. (2013): Transformation of Credit Delivery for the Poor through SHG-Bank Linkage Programme: Retrospect and Prospects. In: Journal of Land and Rural Studies 1 (1): 55-74.

MoLE – Ministry of Labour and Employment, Government of India (2009): National Policy on Skills Development. http://labour.nic.in/upload/uploadfiles/files/Policies/NationalSkillDevelopmentPolicy-Mar09.pdf (May 05, 2017)

Molt, W. (1970): Die Pädagogik von Mahatma Gandhi. Meisenheim: Hain.

Moodithaya, M.S. (2009): Rural Self-Employment Programmes in India: An Appraisal. New Delhi: Manak Publications.

MoRD – Ministry of Rural Development, Government of India (2014): Annual Report 2012-13. New Delhi. http://rural.nic.in/sites/default/files/MoRDEnglish_AR2012_13_0.pdf (October 22, 2017)

MoRD – Ministry of Rural Development, Government of India (2017): Annual Report 2016-17. New Delhi (Chapter 3: Rural Livelihoods: 23-45). http://rural.nic.in/sites/default/files/Annual_Report_2016_17_English_1.pdf (October 5, 2017)

MoT – Ministry of Textiles, Government of India (2010): Handloom Census of India 2009-2010: Third National Census of Handloom Weavers and Issue of Photo Identity Cards to Weavers and Allied Workers. Primary Handloom Census Abstract. New Delhi: National Council of Applied Economic Research. http://handlooms.nic.in/Writereaddata/Handloom%20report.pdf (10.10.2017)

MoT – Ministry of Textiles, Government of India (2017a): Guidelines for Implementation of Comprehensive Handloom Cluster Development Scheme (CHCDS) for Development of Mega Handloom Clusters during the Period from April 2017 to March 2020. New Delhi: Office of the Development Commissioner (Handlooms). http://handlooms.nic.in/writereaddata/1202.pdf (July 23, 2018)

MoT – Ministry of Textiles, Government of India (2017b): Handloom Clusters in Bihar. Press Information Bureau. Released on July 20, 2017 under the ID :168753. http://pib.nic.in/newsite/mberel.aspx?relid=168753 (October 10, 2017)

MSDE – Ministry of Skill Development and Entrepreneurship, Government of India (2015): National Policy for Skill Development and Entrepreneurship. New Delhi. http://www.skilldevelopment.gov.in/assets/images/Skill%20India/policy%20booklet-%20Final.pdf (October 4, 2017)

MSDE – Ministry of Skill Development and Entrepreneurship, Government of India (2016): PMKVY – Pradhan Mantri Kaushal Vikas Yojana Guidelines (2016-2020). http://pmkvyofficial.org/App_Documents/News/PMKVY%20Guidelines%20(2016-2020).pdf (September 18, 2017)

Müller-Mahn, D. (2001): Fellachendörfer. Sozialgeographischer Wandel im ländlichen Ägypten. In: Erdkundliches Wissen 127. Stuttgart: Steiner.

Müller-Mahn, D. (2011): Die Auflösung von Norden und Süden: Neue Raumbilder als Herausforderungen für die Geographische Entwicklungsforschung. In: Gebhardt, H.; Glaser, R.; Radtke, U.; Reuber, P. (eds.): Geographie. Physische und Humangeographie. Heidelberg: Spektrum Akademischer Verlag: 763-783.

Müller-Mahn, D.; Verne, J. (2010): Geographische Entwicklungsforschung: Alte Probleme, neue Perspektive. In: Müller-Mahn, D (ed.): Geographische Rundschau 62 (10): 4-11.

Muniandi, K. (1985): Gandhian Experiments on Education. Gandhigram: Gandhigram Rural Institute.

MYAS – Ministry of Youth Affairs and Sports, Government of India (2014): National Youth Policy 2014. New Delhi. www.rgniyd.gov.in/sites/default/files/pdfs/scheme/nyp_2014.pdf (April 15, 2017)

National Cooperative Union of India (2012): Indian Cooperative Movement 2012: A Statistical Profile. New Delhi. http://www.ncui.coop/pdf/Indian-Cooperative-Movement-a-Profile-2012.pdf (October 22, 2017)

Navale, A.; Sharma, M. (2006): Empowerment: Concept, Evolution and Strategies. In: Verma, R. B. S.; Verma, H. S.; Singh, R. K. (eds.): Empowerment of the Weaker Sections in India: Interface of the Civil Society Organizations and Professional Social Work Institutions. New Delhi: Serials Publications: 64-82.

Navaneetham, K. (2010): Demographic Dividend in India. In: Lakshmanasamy, T. (ed.): Population Dynamics and Human Development: Opportunities and Challenges. New Delhi: Bookwell: 71-80.

NCEUS – National Commission for Enterprises in the Unorganized Sector, Government of India (2008): A Special Program for Marginal and Small Farmers. New Delhi. http://dcmsme.gov.in/Special_Programme_for_Marginal_and_Small_Farmers.pdf (September 5, 2017)

NDTV – New Delhi Television Limited (2014): 10 Big Quotes from PM Narendra Modi's Sydney Speech. Updated: November 17, 2014. https://www.ndtv.com/cheat-sheet/10-big-quotes-from-pm-narendra-modis-sydney-speech-698895 (August 2nd, 2018)

Nelson, P. J. (1995): The World Bank and Non-Governmental Organizations: The Limits of Apolitical Development. Basingstoke: Macmillan.

NICRA – National Innovations on Climate Resilient Agriculture (n.a.): Agriculture Contingency Plan for Respective Districts. http://www.nicra-icar.in/nicrarevised/images/statewiseplans/Bihar/ (November 1, 2017)

NIRD – National Institute of Rural Development, Ministry of Rural Development, Government of India (2011): RSETI: Course Modules for Entrepreneurship Development Training Programmes. 2nd Edition. Hyderabad: NIRD.

NSSO – National Sample Service Office (2004): 61th Round. July 2004-June 2005.

NSSO – National Sample Service Office (2011): 68th Round. July 2011-June 2012.

Palanithurai, G. (2016): Other Official / Governmental Programmes: Government Initiatives for Preparing Youth for the World of Work. In: Pilz, M (ed.): India: Preparation for the World of Work – Education System and School to Work Transition. Wiesbaden: Springer VS: 141-168.

Parker, S. C. (2004): The Economics of Self-Employment and Entrepreneurship. New York: Cambridge University Press.

Planning Commission, Government of India (2008): Eleventh Five Year Plan 2007-2012. Volume I: Inclusive Growth. New Delhi. http://planningcommission.nic.in/plans/planrel/fiveyr/11th/11_v1/11th_vol1.pdf (May 8, 2017)

Planning Commission, Government of India (2009): Percentage of Population below Poverty Line. 18th Round. http://planningcommission.gov.in (February 28, 2017)

Planning Commission, Government of India (2013): Twelfth Five Year Plan, 2012-2017. Volume III: Social Sectors. New Delhi: Sage Publications India. http://planningcommission.gov.in/plans/planrel/12thplan/pdf/12fyp_vol2.pdf (May 8, 2017)

PMKVY – Pradhan Mantri Kaushal Vikas Yoyana (2015): Homepage of PMKVY. http://www.pmkvyofficial.org (June 4, 2018)

Prahalad, C. K. (2014): The Fortune at the Bottom of the Pyramid – Eradicating Poverty through Profits. Revised and Updated 5th Anniversary Edition. Upper Saddle River: Prentice Hall.

Prajapati, K.; Biswas, S. N. (2011): Effect of Entrepreneur Network and Entrepreneur Self-efficacy on Subjective Performance: A Study of Handicraft and Handloom Cluster. In: The Journal of Entrepreneurship 20 (2): 227-247.

Prakasha, V. (1985): Gandhian Basic Education as Programme of Interdisciplinary Instruction at the Elementary Stage: Some Lessons of Experience. Paris: Unesco. (Special UPEL Issue No. 2)

Prasad, J. (2007): Bihar – Dynamics of Development. New Delhi: Mittal Publications.

Rathakrishnan, L. (ed.) (2008): Empowerment of Women through Entrepreneurship. New Delhi: Gyan Publishing House.

Rauch, T. (2009): Entwicklungspolitik: Theorien, Strategien, Instrumente. Das Geographische Seminar. Braunschweig: Westermann.

Renn, J. (2005): Die gemeinsame menschliche Handlungsweise. Das doppelte Übersetzungs-problem des sozialwissenschaftlichen Kulturvergleichs. In: Srubar, I.; Renn, J.; Wenzel, U. (eds.): Kulturen vergleichen. Sozial- und kulturwissenschaftliche Grundlagen und Kontroversen. Wiesbaden: VS Verlag für Sozialwissenschaften: 195-227.

Reynolds, P. D.; Bygrave, W.; Autio, E.; Hay, M. (2002): Global Entrepreneurship Monitor. 2002 Summary Report. Kansas City: Ewin Marion Kauffman Foundation.

Richter, H. (1973): Grundsätze und System der Transkription – IPA (G) PHONAI. 3rd Edition. Tübingen: Niemeyer.

Rigg, J. (2006): Land, Farming, Livelihoods, and Poverty: Rethinking the Links in the Rural South. In: World Development 34 (1): 180–202.

Robson, E.; Bell, S.; Klocker, N. (2007): Conceptualizing Agency in the Lives and Actions of Rural Young People. In: Panelli, R.; Punch, S.; Robson, E. (eds.): Global Perspectives on Rural Childhood and Youth: Young Rural Lives. New York: Routledge: 135-148.

Rooks, G., Sserwanga, A., Frese, M. (2016): Unpacking the Personal Initiative-Performance Relationship: A Mulit-Group Analysis of Innovation by Ugandan Rural and Urban Entrepreneurs. In: Applied Psychology: An International Review 65 (1): 99-131.

Rostow, W. (1960): The Stages of Economic Growth: A Non-Communist Manifesto. Cambridge: Cambridge University Press.

Roy, A. K. (2013): Growth Dynamics and Linkages of Livestock and Dairy Development inthe SSP Command in Gujarat. In: Chandra, K. S.; Babu, V. S.; Nath, P. K. (eds.): Agrarian Crisis in India: The Way Out. New Delhi: Academic Foundation: 283-303.

Roy, U.N.; Saini, J.S. (eds.) (2009): People's Empowerment and Sustainable Rural Development. A Technological Approach. Jaipur. New Delhi: Rawat Publications.

Rubin, H.J.; Rubin, S. I. (1995): Qualitative Interviewing: The Art of Hearing Data. Thousand Oaks: Sage Publications.

RSETI – Rural Self Employment Training Institutes (n.a.): Homepage of RSETI. http://www.rsetimis.org (June 4, 2018)

Sankaranarayanan (2011): Inclusive Education and Sustainable Growth. In: Pillai, L.; Remesh, B. P. (ed.): Bridging the Gap: Essays on Inclusive Development and Education. Los Angeles: Sage: 52-56.

Saraswathi, T. S. (1999): Adult-Child Continuity in India: Is Adolescence a Myth or an Emerging Reality? In: Saraswathi, T. S. (ed.): Culture, Socialization and Human Development: Theory, Research and Applications in India. New Delhi: Sage: 213-232.

Sarka, R.; Sinha, A. (2015): Another Development: Participation, Empowerment and Well-being in Rural India. London. New York. New Delhi: Routledge.

Sarmistha, U. (2015): Rural Handloom Textile Industry in Bihar: A Case of Rural Informal Sector. In: Social Change 45 (1): 107-117.

Schlottmann, A. (2007): Handlungszentrierte Entwicklungsforschung: Das Instrument der Schnittstellenanalyse am Beispiel eines Agroforstprojekts in Tanzania. In: Werlen, B. (ed.): Sozialgeographie alltäglicher Regionalisierungen (3): Ausgangspunkte und Befunde empirischer Forschung. Stuttgart: Steiner: 69-108.

Scholten, B. (2010): India's White Revolution: Operation Flood, Food Aid and Development. London: Tauris Academic Studies.

Scholz, F. (1988): Position und Perspektiven geographischer Entwicklungsforschung. Zehn Jahre Arbeitskreis Entwicklungstheorien. In: Leng, G; Taubmann, W. (eds.): Geographische Entwicklungsforschung im interdisziplinären Dialog, Bremer Beiträge zur Geographie und Raumplanung 18: 9-35.

Scholz, F. (2000): Perspektiven des "Südens" im Zeitalter der Globalisierung. In: Geographische Zeitschrift 88 (1): 1-20.

Scholz, F. (2002): Die Theorie der "fragmentierenden Entwicklung". In: Geographische Rundschau 54 (10): 6-10.

Scholz, F. (2004): Geographische Entwicklungsforschung. Berlin.Stuttgart: Borntraeger.

Schumpeter, J. A. (1934): The Theory of Economic Development: An Inquiry into Profits, Capital, Credit, Interest, and the Business Cycle. Cambridge: Havard University Press.

Schütze, F. (1977): Die Technik des narrativen Interviews in Interaktionsfeldstudien: Dargestellt an einem Projekt zur Erforschung von kommunalen Machtstrukturen. Arbeitsberichte und Materialien Nr. 1. Bielefeld: Universität, Fakultät für Soziologie.

Scoones, I. (1998): Sustainable Rural Livelihoods. A Framework for Analysis. Institute for Development. Studies Working Paper 72. Brighton: IDS.

Scoones, I. (2017): Young People and Agriculture: Implications for Post-Land Reform Zimbabwe. In: Zimbabweland, March 18, 2017. www.zimbabwesituation.com/news/zimsit-m-young-people-and-agriculture-implications-for-post-land-reform-zimbabwe (June, 21, 2017)

Sen, A. (1987): The Standard of Living: The Tanner Lectures, Clare Hall. Cambridge: Cambridge University Press.

Senghaas, D. (1974): Peripherer Kapitalismus: Analysen über Abhängigkeit und Unterentwicklung. Frankfurt am Main: Suhrkamp.

Sengupta, A. K., National Commission for Enterprises in the Unorganised Sector, Government of India (2007): Report on the Conditions of Work and Promotion of Livelihoods in the Unorganised Sector. New Delhi: Dolphin Printo Graphics. http://dcmsme.gov.in/Condition_of_workers_sep_2007.pdf (June 24, 2017)

Shah, A. (2006): Changing Interface between Agriculture and Livestock: A Study of Livelihood Options under Dry Land Farming System in Gujarat. Working Paper no. 170. Ahmedabad: FAO. http://www.fao.org/wairdocs/lead/ae752e/ae752e00.htm (July 24, 2018)

Shanmugam, K.R.; Moorthi, S.S. (2010): Population Growth and Regional Disparities in India. In: Lakshmanasamy, T. (ed.): Population Dynamics and Human Development: Opportunities and Challenges. New Delhi: Bookwell: 289-303.

Simon, D. (2003): Dilemmas of Development and the Environment in a Globalising World: Theory, Policy and Praxis. In: Progress in Development Studies 3 (1). Thousand Oaks: Sage Publications: 5-41.

Singer, A. E. (2006): Business Strategy and Poverty Alleviation. In: Journal of Business Ethics 66 (2-3): 225-231.

Singh, A. (2016): The Process of Social Value Creation. A Multiple-Case Study on Social Entrepreneurship in India. New Delhi: Springer.

Singh, K. M.; Singh, R. K. P.; Jha, A. K.; Meena, M. S. (2010): Dynamics of Livestock Sector in Bihar: A Temporal Analysis. In: Agricultural Situation in India (66), March 15, 2010: MPRA Paper No. 47094: 687-702. http://mpra.ub.uni-muenchen.de/47094/ (October 21, 017)

Sinha, R. R. K. (2009): Dynamics of Land-Caste Relations in India: A Case Study of Bihar. New Delhi: Manak Publications.

Sopam, R. (2017): Rangoli, an All-Women Initiative to Empower Rural Women in Bihar's Madhubani. In: Hindustan Times, July 14, 2017. http://www.hindustantimes.com/india-news/rangoli-an-all-women-initiative-to-empower-rural-women-in-bihar-s-madhubani/story-l714VbejVyMLIxTnehPlAJ.html (October 8, 2017)

Stein, M. (2017): Allgemeine Pädagogik. 3rd Edition. München. Basel: Ernst Reinhardt Verlag.

Sternberg, R.; Bergmann, H. (2003): Global Entrepreneurship Monitor: Länderbericht Deutschland 2012. Köln: Wirtschafts- und Sozialgeographisches Institut, Universität zu Köln. https://www.kfw.de/Download-Center/Konzernthemen/Publikationen-der-ehemaligen-DtA/Pdf-Dokumente-ehemalige-DtA/DtA_GEM.pdf (July 26, 2018)

Tarakumari, P. (2008): Women in Informal Service Sector: A Case Study of Beauticians in Visakhapatnam City. In: Rathakrishnan (ed.): Empowerment of Woman through Entrepreneurship. New Delhi: Gyan Publishing House: 372-387.

Temple, B.; Young, A. (2004): Qualitative Research and Translation Dilemmas. In: Qualitative Research 4 (2): 161-178.

Tröger, S. (2004): Handeln zur Ernährungssicherung im Zeichen gesellschaftlichen Umbruchs. Untersuchungen auf dem Ufipa-Plateau im Südwesten Tansanias. Saarbrücken: Verlag für Entwicklungspolitik.

Trommsdorff, G. (2008): Kultur und Sozialisation. In: Hurrelman, K.; Grundmann, M.; Walper, S. (eds.): Handbuch Sozialisationsforschung. 7th Edition. Weinheim. Basel: Beltz: 229-239.

United Nations, Department of Economic and Social Affairs, Population Division (2011): World Population Prospects: The 2010 Revision. Volume I: Comprehensive Tables. http://www.un.org/en/development/desa/population/publications/pdf/trends/WPP2010/WPP2010_Volume-I_Comprehensive-Tables.pdf (August 2nd, 2018)

Vasanthagopal, R.; Santha, S. (eds.) (2008): Women Entrepreneurship in India. New Delhi: New Century Publishing.

Venkatesh, R. (2014): Learning at Bodhshala: Re-Orienting the School to its Community. Mapusa: Other India Press.

Venkatram, R. (2012): Vocational Education and Training System (VET) in India. In: Pilz, M. (ed.): The Future of Vocational Education and Training in a Changing World. Wiesbaden: Springer VS: 171-178.

Verma, R. B. S. (2006): Empowerment: Concept, Objectives and Strategies. In: Verma, R. B. S.; Verma, H. S.; Singh, R. K. (eds.): Empowerment of the Weaker Sections in India. New Delhi: Serials Publications: 52-63.

Watts, M.; Bohle, H.-G. (1993): The Space of Vulnerability: The Causal Structure of Hunger and Famine. In: Progress in Human Geography 17 (1): 43-67.

Webb, J. W.; Morris, M. H.; Pillay, R. (2013): Microenterprise Growth at the Base of the Pyramid: A Resource-based Perspective. In: Journal of Developmental Entrepreneurship 18 (4) 1350026: 1-20.

Werlen, B. (1997): Gesellschaft, Handlung und Raum: Grundlagen handlungstheoretischer Sozialgeographie. 3rd Edition. Stuttgart: Steiner.

Werlen, B. (1999): Sozialgeographie alltäglicher Regionalisierungen: Zur Ontologie von Gesellschaft und Raum. 2nd Edition. Stuttgart: Steiner.

Werlen, B. (2012): Anthony Giddens. In: Eckardt, F. (ed.): Handbuch Stadtsoziologie. Wiesbaden: Springer VS: 145-166.

Williams, N.; Williams, C. C. (2011): Beyond Necessity Versus Opportunity Entrepreneurship: Some Lessons from English Deprived Urban Neighbourhoods. In: International Entrepreneurship and Management Journal (10):23-40. Published online: June 5, 2011: Springer Science and Business Media.

Witzel, A. (1985): Das problemzentrierte Interview. In: Jüttemann, G. (ed.): Qualitative Forschung in der Psychologie. Weinheim: Beltz: 227-256.

Yunus, M. (1987): Credit for Self-Employment: A Fundamental Human Right. Dhaka: Grameen Bank.

Yunus, M., Moingeon, B., Lehmann-Ortega, L. (2010): Building Social Business Models. Lessons from the Grameen Experience. In: Long Range Planning 43. London: Elsevier: 308-325.

Zinnecker, J. (1982): Jugend '81: Portrait einer Generation. In: Fischer, A.; Fischer R. C.; Fuchs, W.; Zinnecker, J. (eds.): Jugend `81: Lebensentwürfe, Alltagskulturen, Zukunftsbilder. 9. Shell Jugendstudie. 2nd Edition. Opladen: Leske und Budrich: 80-122.

Appendices

Appendix I: Dairy Entrepreneurship Curriculum of Drishtee (Summarized)

Main Training Unit	Main Module	Sub Module	Duration
The cattle shed (creating a healthy environment for the cows)	Construction Hygiene and cleanliness Usage of cow dung		Day 1-6 24 hours
Feeding of the animals	Nutrition (food and water requirements) Feeding of the animals Survival during dry season/drought		Day 7-11 20 hours
Health treatment	Control of diseases Vaccination		Day 12-15 16 hours
Milking	Milking equipment Correct practices		Day 16-18 12 hours
Food storage	Food procurement and food storage Occupational health and safety requirements		Day 19-23 20 hours
Entrepreneurship **5 days** **(20 hours)**	Dairy farming as a self-employment venture (for making money)	Government schemes/lending schemes to the agricultural sector	Day 24-25 8 hours
		Book keeping, project report	
	Market knowledge	Market prices	Day 26 4 hours
		Classification of milk products	
	Customer relations	Establishing customer relations	Day 27 4 hours
		Customer needs, and how to meet them	
		Finding market gaps	
	Cost calculation	Market prices, finding the right buyers	Day 28 4 hours
		New technologies	
		Finding the break-even point	
Safety standards	Safety standards and regulations		Day 29-32 16 hours

Source: Author's own summary of an original Hindi version of the Dairy Entrepreneurship Curriculum of Drishtee (date of publication unspecified; emphasis of the entrepreneurship unit added), and obtained personally from Drishtee in Bhagalpur, Bihar.
The summary is based on an English translation made by an independent interpreter.

Appendix II: RSETI Dairy Farming Course Module

DAIRY FARMING

Day	Session	Subject
01	I	Registration & Inauguration
	II	About the Institute, rules & regulations of training/institute
	III	Micro lab – Ice breaking exercise
	IV	Achievement Motivation - confidence building
02	I	Entrepreneurial competencies - importance, explanation with examples, case study for identification of different competencies
	II	Dairy farming as a sustainable self employment venture - prospects
	III	Dairy farming, breeds of cows and buffaloes, up gradation of cattle by cross breeding, selection of animals
	IV	Dairy farming - methodology of correct practices, misconceptions
03	I	Nutrition & feeding of dairy animals, preparation of feeds & use of azola
	II	Fodder crops - description, cultivation aspects
	III	Calf rearing & calf management practices for production of a healthy cow/buffaloes
	IV	Cattle shed - construction, importance of hygiene & cleanliness - management practices
	Post evening	Tower building - eradicating dependency syndrome
04	I	Dairy animals - important diseases & their control, vaccination
	II	Artificial insemination - procedural details, management of animals in pregnancy
	III	Production of a clean milk - practices, milk products
	IV	Banking-deposits & advances, lending schemes to agricultural sector, Government schemes
05	I	Field visit for interface with successful dairy farmers
	II	Use of cow dung & urine for preparation of Farm Yard Manure (FYM), bio gas plant, compost pit preparation
	III	Economics of a Dairy unit - preparation of project report
	IV	Time Management
06	I	Insurance
	II	Problem solving-explanation through case studies and exercises, creativity - creative thinking
	III	Renewable Energy, an appropriate alternative - description, scope
	IV	Feedback & Valedictory

Space for updations/additions:

Source: NIRD 2011, 69

Appendix III: RSETI Advanced Dairy Management Course Module

ADVANCED DAIRY MANAGEMENT

Day	Session	Subject
01	I	Registration & Inauguration
	II	About the Institute, rules & regulations of the training/institute
	III & IV	Micro lab - Ice breaking exercise
02	I	Achievement Motivation - Confidence building
	II	Why Self employment - Advantages over wage employment, Entrepreneurship Development - What, Why & How? - (introduction)
	III & IV	Entrepreneurial competencies - Importance, explanation with examples, case study for identification of different competencies
03	I	Briefing about BAIF & its Projects
	II	Briefing about Ksheeradhara programme & its objective
	III & IV	Importance of livestock in Indian Scenario, particularly cows & buffaloes, Identification of different breeds and blood level of animals
04	I to IV	Description of breeds - Cows and buffaloes, exercise for assessing blood level of animals
05	I	Tower Building - Decision making & eradicating dependency syndrome
	II to IV	Breeding of cows and buffaloes, female reproductive system - Description
06	I & II	Effective communication skills
	III & IV	Breeding policy of the state, Casting of animals
07	I	Problem Solving - Explanation through case studies and exercises
	II to IV	Systems of mating/breeding & its importance Handling of specimen of female genital organ, fodder species
08	I & II	Handling specimen of female reproductive organs
	III & IV	Palpation of female reproductive system
09	I	Experience sharing - Interaction with successful entrepreneur
	II to IV	Symptoms of heat, Handling of female reproductive system
10	I	Business game - Boat Building Exercise
	II to IV	Third eye - Internalization of competencies Hormonal regulation of estrus cycle, Handling of Artificial Insemination equipment, semen, LN2
11	I to IV	Methods of breeding, Artificial Insemination technique
12	I to IV	Semen doses, Thawing
13	I & II	Market survey - Theory
	III & IV	Artificial Insemination Guns and their description & uses, Artificial insemination
14	I to IV	Description of different sheaths Pregnancy diagnosis
15	I	Marketing management - 4 Ps of marketing, managing the customers
	II to IV	Liquid nitrogen & its importance
	Post evening	Risk taking and goal setting - Ring Toss exercise
16	I & II	Handling of Artificial Insemination equipment, semen, and LN2Identification of different feeds and fodder
	III & IV	Pregnancy diagnosis in cows & buffaloes
17	I	Time Management
	II to IV	Handling of genital organs of cows & buffaloes
18	I to IV	Heat detection and Artificial Insemination
19	I & II	Visit to successful entrepreneur units

(Continued on next page)

Day	Session	Subject
	III & IV	Techniques of artificial insemination
	Post evening	Final evaluation test
20	I	Costing, pricing - Fixed Cost - Variable Cost, Break even point etc.
	II	Business plan/project report preparation - Practical
	III	Maintenance of records & book keeping - Methodology
	IV	Human Relations - Importance, principles & methodology
21	I & II	Banking - Deposits & advances, lending schemes/Government schemes
	III	Launching formalities-Steps in launching of an enterprise, pitfalls and their control
	IV	Feedback/Valedictory

Space for updations/additions:

Source: NIRD 2011, 71-72

Appendix IV: RSETI Inland Fisheries Course Module

IIIIIIIIIII ▬▬▬ Course Module 2010 ▬▬▬IIIIIIIIIII

PISCICULTURE (INLAND FISHERIES)

Day	Session	Subject
01	I	Registration & Inauguration, about the Institute, rules & regulations of training/institute
	II	Micro lab - Ice breaking exercise
	III	Achievement Motivation - Confidence building
	IV	Entrepreneurial competencies
02	I	Inland Fisheries - Description & scope
	II	Fish culture - Cultivable species - Description
	III	Construction of Pond, types & Management, weed control
	IV	Fish seed culture
03	I & II	Breeding methodology of Major fish species - Indian major carps - description, methods of culture
	III	Common Diseases of fish & their management
	IV	Fresh water prawn culture - Description, methodology
	Post evening	Tower building - Eradicating dependency syndrome
04	I	Fresh water prawn culture - Economics & viability
	II	Prawn - Mono culture & poly culture
	III	Prawn culture in brackish water
	IV	Marketing management
05	I & II	Visit to fishery unit for interface with successful entrepreneur
	III & IV	Economics of fish farming - Preparation of project report
06	I	Time management
	II	Problem solving- Explanation through case studies and exercises, creativity-Creative thinking
	III	Banking - Lending schemes to Agricultural Sector
	IV	Feedback & Valedictory

Space for updations/additions:

▬ 78 ▬▬▬▬▬▬▬▬▬▬▬▬▬▬▬▬ Rural Self Employment Training Institute ▬▬▬

Source: NIRD 2011, 78

Appendix V: RSETI Beauty Parlour Management Course Module

‖‖‖‖‖‖‖‖‖‖ ▬▬▬ Course Module 2010 ▬▬▬ ‖‖‖‖‖‖‖‖‖

BEAUTY PARLOUR MANAGEMENT

Day	Session	Subject
01	I	Registration & Inauguration
	II	About the Institute, rules & regulations of training/institute
	III & IV	Micro lab-Ice breaking exercise
02	I	Achievement Motivation-Confidence building
	II	Why self employment-Advantages over wage employment, Entrepreneurship Development - What, Why & How?-(introduction)
	III & IV	Entrepreneurial competencies - Importance, explanation with examples, case study for identification of different competencies
03	I	Problem solving-explanation through case studies and exercises, Creativity - Creative thinking
	II	Time management
	III & IV	Risk taking and Goal setting - Ring Toss exercise
04	I	The concept of Beautification in women - what, why& how?
	II to IV	Threading & Eye brow shaping - Theory, demonstration & practical
05	I to IV	Waxing - Theory, demonstration and Practical
06	I to IV	Manicure & Pedicure - Theory, demonstration and Practical
07	I to IV	Bleaching of face (cream method) - Theory, demonstration and Practical
08	I to IV	Herbal cream facial - Theory, demonstration and Practical
09	I to IV	Herbal fruits & vegetable facial - Theory, demonstration and Practical
10	I	Business Game - Boat Building Exercise
	II to IV	Galvanic high frequency vat removal - Theory, demonstration and practical
11	I to IV	Aroma Therapy & Acne Treatment - Facial
12	I	Effective communication skills
	II to IV	Hair cutting - Theory, demonstration and practical (Adult)
13	I	Experience sharing - Interaction with successful entrepreneur
	II to IV	Hair cutting - Theory, demonstration and practical (Children)
14	I to III	Advanced Hair cuttings - Theory, demonstration and practical
	IV	Market Survey - Theory
15	I to IV	Market Survey - Collection of information and field visits
16	I & II	Market survey - Report writing, presentation, group discussion & analysis
	III & IV	Hair massage & Body massage - Theory, demonstration and practical
	Post evening	Mid term evaluation test
17	I to IV	Hair straightening (chemical) cum ironing - Theory, demonstration & practical
18	I to IV	Perming - Theory, demonstration and practical
19	I to III	Henna for hair - Theory, demonstration and practical
	IV	Tower building - Eradicating dependency syndrome
20	I to III	Hair colour, hair dye highlights - Theory, demonstration and practical
	IV	Marketing management - 4Ps of marketing, managing the customers
21	I to IV	Spa Treatment - Theory, demonstration and practical
22	I to IV	Bridal Mehandi - Preparation, designing and application - Theory, demonstration and practical
23	I to IV	Make-up & Dressing - Casual, Day, Night, Waterproof - Theory, demonstration and practical

(Continued on next page)

Day	Session	Subject
24	I to IV	Make-up & Dressing - Western & traditional - Theory, demonstration and practical
25	I to IV	Bridal hair Style - Theory, demonstration and practical
26	I to IV	Hair Style using machines - Theory, demonstration and practical
27	I & II	Visit to Beauty Parlours of successful entrepreneurs.
	III & IV	Hair Style by using machines - Theory, demonstration and practical
28	I to IV	Herbal oil/ face pack preparation for different types of skins - Theory, demonstration and practical
	Post evening	Final evaluation test
29	I	Costing, pricing - Fixed cost, variable cost, breakeven point etc.
	II	Business plan/project report preparation
	III & IV	Banking- Deposits & advances, lending schemes/Government schemes
30	I	Human Relations
	II	Maintenance of records & book keeping - Methodology
	III	Launching formalities - Steps in launching of an enterprisePitfalls and their control
	IV	Feedback & Valedictory

Space for updations/additions:

Source: NIRD 2011, 118-119

214

Appendix VI: Interview Guide for Entrepreneurs, who Completed an Entrepreneurship Training

Part I: Personal background

How old are you? Male/female?
In which village/city/block/district do you live? Where did you grow up?
Are you married? How many children do you have, and how old are they?
How many of them go to school?
How many people live in your household? How many of them are already in work?
What kind of work do they do?
What is your father's employment? What is your mother's employment?
How many brothers/sisters do you have?
Do you come from a BPL-family?
How many years of schooling did you complete?
Did you complete, or are you pursuing, a BA/MA?
In which trade do you work? How long have you worked in this trade?
What exactly do you do for work? What do you grow/sell?
Do you own or lease any land? If so, how many acres?
Do you have hired workers? If so, how many?
Did you take any other entrepreneurship training or vocational education and training course before taking this course at Drishtee/RSETI/...?
How did you acquire knowledge in the field of your profession before taking this course at Drishtee/RSETI/...?

Part II: Evaluation of the training course

When did you undertake the entrepreneurship training at Drishtee/RSETI/...?
In what subject did you undertake the entrepreneurship training? How long did the training last?
What motivated you to do this training course?
Have you been able to increase your business afterwards?
Were you able to utilize the training inputs in your work?
How would you rate this course between A (very good) and D (not satisfied)?
What would you like to improve about the training?

Part III: Gaining access to productive livelihoods

1. Effects on human capital

Did this training course motivate you to undertake any other entrepreneurship training/ vocational training and education, or to pursue school/university education? If so, please describe what you did and why.

- Entrepreneurship training/skill development courses completed since the completion of the Drishtee/RSETI/…-course
- Further school/university education completed since the completion of the Drishtee/RSETI/…-course

Did you hire any workers after taking this training course?

2. Effects on financial capital

How did your financial situation change after taking this training course?

- Did you benefit from an increase of monthly income?
- Did you receive a credit from any commercial bank?
- From what time did you receive bank credits? How much? Did you receive it through a SHG/farmers' club?

3. Effects on natural capital (if applicable)

How did your natural capital assets increase after taking this training course?
Did the size of your farm land increase after taking this training course?
Did you change your cultivation method after taking this training course? How do you benefit from it?

Did you increase the number of cows/buffalo after taking this training course?

4. Effects on physical capital

How did the training course affect the following of your physical capital assets?

- Availability of raw materials
- Your means of production (e.g. sewing machines)
- Availability of good quality seeds (for farmers)
- Means of transportation
- Means of communication
- Supporting infrastructure, such as access to education and health facilities, electricity, water, internet, banking services, rural extension services

5. Effects on social capital

Did you become organized in a SHG/farmers' club after taking this training course? How is this organization supporting you in your business activity?

Did you become a member of any other organization or network? Why did you decide to join it? E.g. if you joined one of the following organizations:

- Panchayati Raj Institution / Political party
- NGO
- Trade union/association
- Other (please list)

Part IV: Structural conditions

1. Impact of the socialization agents:

Did your training provider assist you afterwards? How did you benefit from this assistance?
Did the training provider assist you in building up social networks (e.g. by helping you to become a member in SHG/farmers' club)?
How is your family supporting you?

2. Structural limitations:

What obstacles do you face regarding your business activity?
Do you believe that you can actively contribute to change these conditions in your favour?

Part V: Aspirations

What goals do you have for your self-employed activity?
- Increasing your income
- Increasing your number of cows/buffalo/size of land/number of products/...
- Opening new business ventures
- Employing workers
- Others (please list)

What limitations are you confronted with in order to achieve these business goals?
Do you want to pursue further entrepreneurship training? Where? In what subjects?
Do you have any career aspiration besides your self-employed activity?
Do you have any other educational goals which you would like to achieve?
What personal limitations are you confronted with in order to achieve your professional and educational goals?

Thank you very much for your time!

Appendix VII: Interview Guide for Training Providers

For M: Questions for training managers

For T: Questions for various subject teachers

For M/T: Questions for both training managers and subject teachers

For M: What courses do you offer?

- What kinds of trades do you offer?
- How long are the courses?
- What is your average number of students?
- What share of BPL-students are found in the different training courses?
- What kind of certificate do the students receive (upon completion of the course)?

For T: What training course is this?

- How long does it last?
- How many students sign up for it?
- What is the share of BPL-students in your class?
- What kind of certificate do the students receive?

For M: What is your intake capacity? How many applications do you get? How do you select your students?

For M/T: What is the percentage of graduates who manage to get self-employed afterwards? In which trade(s) do you have the most success? (Settlement-rate)

For M/T: What are the main difficulties that your graduates are confronted with in becoming self-employed? Are there differences between male and female graduates?

For M/T: How long do you keep in touch with your graduates afterwards?

For M/T: Please explain how you support your graduates to utilize the training inputs afterwards.

- Do you give them support in getting connected with the market place?
- Support in building strategic social networks?
- Financial support?

For M/T: How well are you connected with the local community?

- Connections to community-based organizations, such as farmer organizations, SHGs, etc.?
- Connections to government institutions/ Panchayati Raj Institutions?
- Connections to NGOs?

For M/T: What other connections do you have? What kinds of connections are these?

- Connections to industry?
- Connections to commercial banks?
- Connections to other training providers?
- Connections to National Skill Development Corporation/ trade unions?

Appendix VIII: Guide for the Narrative Interviews with Village Heads/ Elders

- Socio-economic information about village members: e.g. caste groups, percentage of BPL-families, farming activities, other professions.

- Information about village infrastructure: pre-schools, schools, colleges, training institutions, health care, roads, public transportation, houses, industry, toilet situation, availability of electricity, Internet.

- Information about agricultural-climatic conditions, exposure to drought, floods, other dangerous events.

- Information about the village development: involvement of government, NGOs, community-based organizations, farmers' clubs, SHGs, banks, other agents.

Appendix IX: Guide for Individual and Group Interviews with Trainees

Part I: Personal background

How old are you? Male/female?

In which village/city/block/district do you live? Where did you grow up?

Are you married? How many children do you have and how old are they?

How many people live in your household? How many of them are in work?

What kind of work do they do?

What is your father's employment? What is your mother's employment?

How many brothers/sisters do you have?

Do you come from a Below-the-Poverty-Line family?

How many years of schooling did you complete?

Did you compete, or are you pursuing, a BA/MA?
What is your current profession?

Are you self-employed? In which area(s)?

Did you take any entrepreneurship training or vocational education and training course before you took the present training course?

Part II: Aspirations

What motivated you to do this course?

What is your career aspiration?

Do you have any educational goals which you would like to achieve?

What personal limitations are you confronted with, in order to achieve these goals?

Thank you very much for your time!

Appendix X: The sustainable livelihoods framework of the DFID (UK)

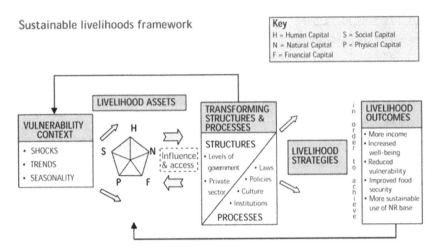

Source: Department for International Development UK (1999): Sustainable Livelihood Guidance
Sheet. DFID, London.
Image taken from: http://practicalaction.org/images/sl-framework-colour.gif (September 1st, 2015)

Printed in the United States
By Bookmasters